Machine Learning: Foundations, Methodologies, and Applications

Series Editors

Books published in this series focus on the theory and computational foundations, advanced methodologies and practical applications of machine learning, ideally combining mathematically rigorous treatments of a contemporary topics in machine learning with specific illustrations in relevant algorithm designs and demonstrations in real-world applications. The intended readership includes research students and researchers in computer science, computer engineering, electrical engineering, data science, and related areas seeking a convenient medium to track the progresses made in the foundations, methodologies, and applications of machine learning.

Topics considered include all areas of machine learning, including but not limited to:

- Decision tree
- Artificial neural networks
- Kernel learning
- Bayesian learning
- Ensemble methods
- Dimension reduction and metric learning
- Reinforcement learning
- Meta learning and learning to learn
- Imitation learning
- Computational learning theory
- Probabilistic graphical models
- Transfer learning
- Multi-view and multi-task learning
- Agents and Multi-Agent Systems
- Graph neural networks
- Generative adversarial networks
- Federated learning
- Large Language Models
- Multimodal Learning
- Transformer/Diffusion Models
- Generative Artificial Intelligence
- Bio-inspired Learning Models
- Embodied AI
- Explainable AI and Ethics

This series includes monographs, introductory and advanced textbooks, and state-of-the-art collections. Furthermore, it supports Open Access publication mode.

Xiangjie Kong • Lingyun Wang •
Mengmeng Wang • Guojiang Shen

Cross-device Federated Recommendation

Privacy-Preserving Personalization

Springer

Xiangjie Kong
College of Computer Science
and Technology
Zhejiang University of Technology
Hangzhou, China

Lingyun Wang
College of Computer Science
and Technology
Zhejiang University of Technology
Hangzhou, China

Mengmeng Wang
College of Computer Science
and Technology
Zhejiang University of Technology
Hangzhou, China

Guojiang Shen
College of Computer Science
and Technology
Zhejiang University of Technology
Hangzhou, China

ISSN 2730-9908 ISSN 2730-9916 (electronic)
Machine Learning: Foundations, Methodologies, and Applications
ISBN 978-981-96-3211-4 ISBN 978-981-96-3212-1 (eBook)
https://doi.org/10.1007/978-981-96-3212-1

This Springer imprint is published by the registered company Springer Nature Singapore Pte Ltd.
The registered company address is: 152 Beach Road, #21-01/04 Gateway East, Singapore 189721, Singapore

Dedicated to our families for their support during the preparation of this book and to all our students for their ideas, motivations, and patience.

Preface

In the last decade, we have seen an exponential increase in the use of recommender systems across various platforms, from e-commerce websites to streaming services. These systems play a crucial role in shaping user experiences by providing personalized content and suggestions tailored to individual preferences. However, with the advent of cross-device usage and the increasing importance of user privacy, new challenges have emerged in the field of recommender systems.

This book, *Cross-device Federated Recommendation: Privacy-Preserving Personalization*, is born out of the need to address these challenges. The aim is to explore the intersection of federated learning and recommender systems, focusing on how we can leverage federated learning to enhance recommendations across multiple devices while ensuring user privacy. Federated learning, a new paradigm in machine learning, allows models to be trained across decentralized devices without sharing raw data, offering a promising solution to privacy concerns.

Each chapter of the book is structured to stand alone, allowing readers to focus on specific areas of interest according to their needs. Chapter 1 introduces federated learning, a method that addresses data privacy, security, and transmission cost issues by keeping data local and sharing only model updates. It outlines the limitations of traditional centralized systems and presents main modes within federated learning. Additionally, it offers an introduction to recommender systems and federated recommendation systems, highlighting the key challenges of data heterogeneity, communication overhead, and privacy concerns. Chapter 2 focuses various learning paradigms in cross-device federated recommendation systems, covering statistical machine learning, deep learning, meta learning, and reinforcement learning. It stresses the importance of efficient aggregation and advanced privacy-preserving methods to manage varied data and model structures. Besides, it provides a comparative analysis of different approaches, revealing their strengths and limitations, and offers insights for advancing federated recommendation systems. Chapter 3 delves into the privacy-preserving mechanisms that are essential in federated learning environments. It covers the implementation of differential privacy, secure multi-party computation, and homomorphic encryption in the context of recommender systems, providing insights into how these techniques are applied in federated recommendation systems to ensure user privacy. Chapter 4 explores federated issues in cross-device federated recommendation systems. It begins by introducing communication-efficient optimization techniques such as knowledge distillation, parameter compression, edge computing, and client clustering. Furthermore, it categorizes existing efforts of fairness into client-level and group-level approaches, offering a structured overview of the current landscape of fairness in federated learning. Chapter 5 concludes the book by exploring future directions and potential research areas in cross-device federated recommendation systems. It discusses emerging trends and also presents an outlook on the challenges that lie ahead and offers insights into the potential solutions that could shape the future of this field.

This book is designed to serve a wide audience, including academic researchers, industry professionals, and students interested in the field of recommender systems and privacy-preserving technologies. It can be used as a reference for both theoretical exploration and practical implementation of cross-device federated recommendation systems.

We sincerely hope that you will find this book both informative and inspiring as we continue to explore the future of privacy-preserving recommender systems together.

Finally, we extend our deepest gratitude to our colleagues and students who contributed their insights and feedback during the writing process. Special thanks go to our families for their unwavering support and understanding throughout this endeavor.

Hangzhou, China
August, 2024

Xiangjie Kong
Lingyun Wang
Mengmeng Wang
Guojiang Shen

Acknowledgments

This book is a product of our research journey on cross-device federated recommendation, particularly within the rapidly evolving domain of federated learning. We extend our heartfelt gratitude to the colleagues and friends whose contributions, feedback, suggestions, and mentorship have been invaluable along the way. While enumerating each name would result in an extensive list, we wish to collectively express our thanks.

We are deeply appreciative of the members of our research teams at Zhejiang University of Technology. Our students and academic peers have been pivotal in our research endeavors, contributing through numerous collaborative sessions and insightful dialogues that have laid the groundwork for this publication. Specific acknowledgments go to Kaiyu Zhang for significant contributions to Chaps. 3, 4, and 5; and Zhenpeng Wu for their substantial input into Chaps. 1, 2, and 5.

Our appreciation also extends to the dedicated staff at Springer, notably Dr. Celine Lanlan Chang, for their unwavering patience and continuous professional support throughout the publishing process.

Besides, this work was supported in part by the National Natural Science Foundation of China under Grant 62072409 and Grant 62073295.

Lastly, yet importantly, we owe a debt of gratitude to our families for their steadfast and selfless support, which has been instrumental in making this endeavor possible.

Competing Interests The authors have no competing interests to declare that are relevant to the content of this manuscript.

Contents

Chapter 1
Introduction

Abstract This chapter introduces federated learning (FL), a distributed machine learning approach designed to address data privacy, security, and processing challenges in the era of big data and AI. FL allows model training across multiple devices or organizations without sharing raw data, thus enhancing privacy by only exchanging model updates. The chapter covers three FL modes: Horizontal Federated Learning (HFL), Vertical Federated Learning (VFL), and Transfer Federated Learning (TFL). It also explores recommender systems (RecSys), focusing on traditional methods like collaborative filtering and advanced techniques, such as graph-based approaches. Finally, the chapter discusses Federated Recommender Systems (FedRec), which integrate FL and recommender systems to improve privacy-preserving recommendation services while addressing challenges like data heterogeneity and communication overhead.

1.1 Federated Learning

Cross-device federated recommendation represents a significant intersection between recommendation systems and federated learning, playing a crucial role in the rapidly evolving landscape of device-centric artificial intelligence. As smart devices become increasingly ubiquitous and interconnected, the demand for personalized recommendations continues to grow. However, the central challenge lies in how to manage the vast amounts of data distributed across diverse devices while ensuring the preservation of user privacy. Federated learning addresses this challenge by enabling the training of models directly on devices, eliminating the need to transfer sensitive user data to centralized servers.

This decentralized learning paradigm not only enhances privacy protection but also leverages the heterogeneous data distribution across different devices, thereby improving model generalization. In the era of smart devices—ranging from smartphones to Internet of Things (IoT) systems—federated learning is pivotal in enabling cross-device personalized recommendations while ensuring data security and regulatory compliance. Therefore, understanding the principles of

X. Kong et al., *Cross-device Federated Recommendation*, Machine Learning: Foundations, Methodologies, and Applications,
https://doi.org/10.1007/978-981-96-3212-1_1

federated learning is fundamental to mastering the concept of cross-device federated recommendation, as it forms the theoretical and technical foundation for building such systems.

In this section, we will provide a comprehensive overview of the background and fundamental concepts of federated learning (FL), exploring its significance in the era of big data and artificial intelligence. We will explain how FL addresses the challenges associated with traditional centralized data processing methods, such as data privacy issues, security risks, high data transmission costs, and processing delays. Additionally, we will introduce the three primary collaborative modes of FL: Horizontal Federated Learning (HFL), Vertical Federated Learning (VFL), and Transfer Federated Learning (TFL). Following this, we will discuss the implementation and advantages of FL in cross-device and cross-silo scenarios. Finally, we will examine the various challenges that FL faces in practical applications, including data heterogeneity, communication overhead, privacy and security concerns, and issues related to computational resources and energy consumption. By understanding these foundational concepts and challenges, readers will gain a comprehensive understanding of FL's flexibility and efficiency in different data distribution and application contexts.

1.1.1 Background and Concepts

The need for data privacy, security issues, and the demand for distributed data processing in the era of big data have driven the emergence of federated learning (FL). In the rapid development of big data and artificial intelligence, data has become a crucial resource driving the advancement of machine learning models. However, traditional centralized data processing methods face severe privacy and security risks, high data transmission costs, and processing delays.

With the proliferation of digitalization and smart devices, various devices generate massive amounts of data daily. This data holds immense potential value but presents significant challenges when processed in a centralized manner. First, privacy protection has become an indispensable concern. The collection and storage of large-scale datasets, especially those containing personal sensitive information, face risks of unauthorized access and misuse. In recent years, a series of major data breaches have heightened public awareness of data privacy, prompting the introduction and enforcement of stricter data protection regulations. Individuals, organizations, and society as a whole are strengthening the protection of data security and privacy. A notable example is the General Data Protection Regulation (GDPR) [18], implemented by the European Union on May 25, 2018. The GDPR aims to protect users' personal privacy and data security by requiring operators to clearly communicate the content of user agreements and prohibiting deceptive or coercive practices to waive privacy requirements. Furthermore, operators are not allowed to train models without explicit user permission and must also allow users to delete their private data. Similarly, since 2017, China's Cyber Security Law of the

People's Republic of China [26] and the General Principles of the Civil Law of the People's Republic of China [10] have explicitly stated that network operators must not disclose, tamper with, or destroy the personal information they collect. When conducting data transactions with third parties, it is necessary to ensure that the contract clearly specifies the scope of data being traded and the obligations for data protection. These regulations force companies to seek new data processing methods to ensure compliance and protect user privacy.

Moreover, traditional centralized data processing methods require transferring all data to a central server for processing, which is not only costly but can also lead to network congestion and delays. In today's landscape, where IoT devices and mobile devices are widely used, the data generated by these devices is distributed globally. Efficiently processing this dispersed data has become a significant challenge. For example, smartphones, wearable devices, and smart home devices generate vast amounts of user behavior and environmental data daily. Collecting and processing this data centrally is not only technically challenging but also economically unsustainable.

In response to these challenges, federated learning [22] has emerged as a novel distributed machine learning method. Federated learning (FL) is an innovative distributed machine learning method that enables collaborative model training without sharing raw data. This approach emerged to address the challenges associated with traditional centralized data processing methods, including privacy concerns, security risks, and high transmission costs. Federated learning allows participants to train models locally on their devices and then upload only the model parameters or gradients to a central server for aggregation, significantly reducing data transmission volumes and effectively protecting data privacy.

One of the key features of federated learning is its decentralized approach to data processing, which not only complies with stringent data privacy regulations but also enables efficient data handling in distributed environments. The advantage of federated learning lies in its ability to facilitate cross-organization and cross-regional machine learning model training while preserving data privacy. For example, in the medical field, the sensitivity of patient data and privacy protection requirements make it difficult for medical institutions to share data for centralized processing. However, through federated learning, these institutions can collaboratively train a disease prediction model without sharing patient data, thereby enhancing the accuracy and generalizability of diagnoses.

A classic and representative implementation of federated learning is the federated averaging algorithm (FedAvg). The fundamental idea of FedAvg is to distribute the model training tasks to multiple participants, each of whom trains the model independently on their local devices and then sends the trained model parameters to a central server for aggregation. This iterative process gradually optimizes the

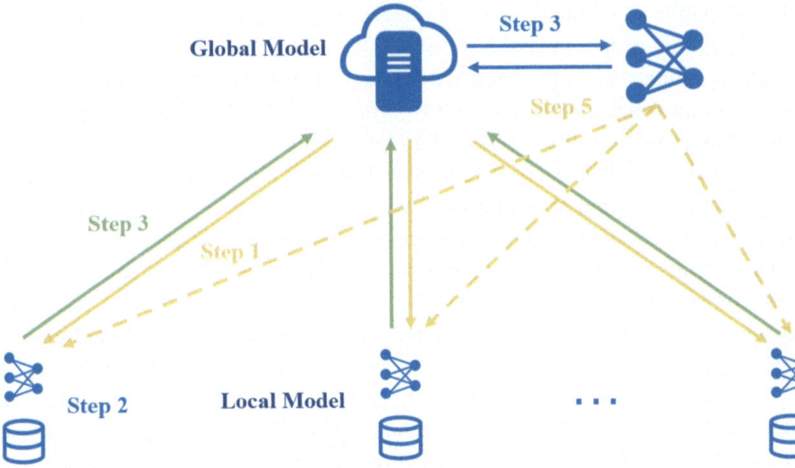

Fig. 1.1 A schematic diagram of federated learning

global model. The process of federated learning can be referenced in the schematic diagram (Fig. 1.1). The workflow of FedAvg is as follows:

- **Step 1. Initialization of the Global Model.** The central server initializes a global model and distributes copies to all participants. This global model serves as the baseline, upon which each participant conducts local training.
- **Step 2. Local Model Training.** Upon receiving the global model, each participant uses local data to train the model for several rounds, updating the model parameters. Since local data may be heterogeneous (differently distributed), the trained model parameters can vary.
- **Step 3. Uploading Model Parameters.** After local training, each participant uploads the updated model parameters to the central server. This step involves transmitting only the model parameters, not the raw data, thereby significantly reducing data transmission volumes.
- **Step 4. Aggregating Model Parameters.** The central server aggregates the received model parameters using a weighted average method to generate a new global model. The weights are typically determined based on the amount of data each participant has, ensuring that participants with more data have a greater influence on the aggregation.
- **Step 5. Distributing the Aggregated Model.** The central server redistributes the aggregated global model to all participants for the next round of local training.

This iterative process continues until the model reaches the desired accuracy or convergence. By combining local training and central aggregation, the FedAvg algorithm enhances model training efficiency while protecting participants' data privacy, making it suitable for various applications requiring distributed data processing.

For instance, in the financial sector, different banks can collaborate using the FedAvg algorithm to develop fraud detection models without sharing sensitive customer information, thereby enhancing the security of financial systems while maintaining privacy. Overall, federated learning and its representative algorithm, FedAvg, provide an effective solution for distributed data processing and privacy protection, promoting the widespread application and development of artificial intelligence technologies across various fields.

In the context of recommendation systems, federated learning plays a crucial role by enabling collaborative filtering and model training across multiple devices and platforms without compromising user privacy. Traditional recommendation systems often rely on centralized data collection, posing significant privacy risks. Federated learning mitigates these risks by ensuring that user data remains on their local devices, thus adhering to privacy regulations while still leveraging the collective data insights for accurate recommendations.

For instance, in a music streaming service, federated learning can help generate personalized song recommendations by training a model across the data from multiple users' listening habits without centralizing their data. Each user's device trains a local model based on their listening history, and only the model updates are shared with the central server. This way, the service can provide accurate recommendations while ensuring that the users' listening data remains private.

Similarly, in e-commerce platforms, federated learning can improve product recommendations by leveraging the shopping behavior data of users across different regions. Each user's device trains a local model with their shopping history, and the aggregated model updates help the central server to refine the recommendation algorithm. This approach enhances the recommendation accuracy without compromising user privacy, leading to a better shopping experience and increased user satisfaction.

In summary, federated learning represents a significant advancement in the field of machine learning, particularly in applications requiring stringent data privacy and security measures. Its decentralized approach and efficient handling of distributed data make it a cornerstone technology for the future of intelligent systems.

1.1.2 Horizontal, Vertical, and Transfer Federation

Federated learning (FL) encompasses three primary collaborative modes: Horizontal Federated Learning (HFL) [22], Vertical Federated Learning (VFL), and Transfer Federated Learning (TFL) [32]. These modes address different data distribution scenarios and application needs, providing flexible and efficient data processing and model training methods while ensuring data privacy across organizations and regions. To effectively leverage federated learning for cross-device recommendation systems, it is crucial to understand the distinct characteristics and advantages of these three modes.

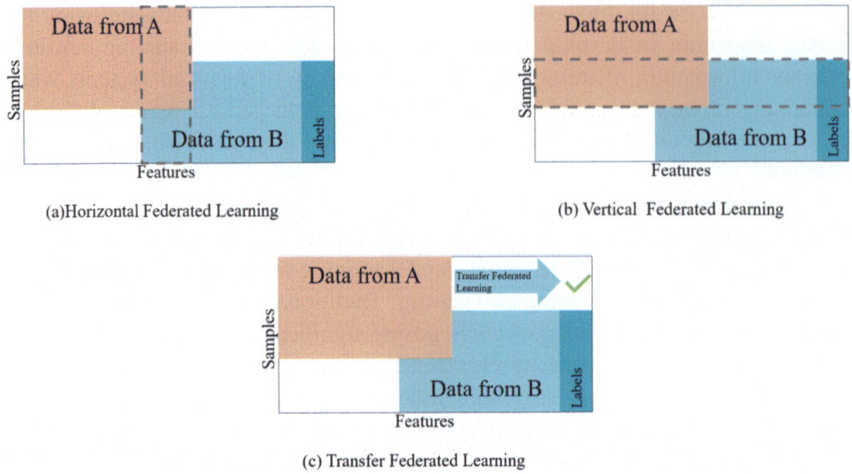

Fig. 1.2 Three primary collaborative modes of federated learning

Horizontal Federated Learning is suitable when datasets from different parties have highly overlapping user features but little overlap in user samples. Specifically, HFL involves horizontally partitioning datasets by user dimension, extracting data segments where user features are the same but user samples differ for training. This method effectively increases the total number of training samples, thus improving model accuracy. For instance, two banks in different regions may collect similar financial data from different user groups. By employing HFL, these banks can collaboratively train models without sharing raw data (Fig. 1.2).

The advantages of HFL include its ability to significantly enhance data privacy protection since data remains local, greatly reducing the risk of privacy breaches. Additionally, HFL offers excellent scalability, capable of handling a large number of participants, each contributing to model improvement without sharing raw data. By combining data from multiple sources, HFL can increase the number of training samples, thus improving model performance. However, HFL also faces challenges such as data heterogeneity and communication overhead. Differences in data distributions across parties can affect the performance of the aggregated model, while frequent transmission of model parameters can lead to significant network overhead. To address these issues, common solutions include homomorphic encryption, differential privacy (DP), and secure aggregation techniques to ensure data security during gradient exchanges in HFL.

In cross-device recommendation systems, HFL can be applied to scenarios where different service providers have overlapping user features but distinct user bases. For example, various music streaming services could use HFL to collaboratively improve their recommendation algorithms by combining data on user listening habits, thereby providing more accurate song recommendations without compromising user privacy. Similarly, fitness app developers can share insights derived from user workout data to enhance personalized fitness plans while maintaining individual data privacy.

Vertical Federated Learning (VFL) is suitable when datasets from different parties have highly overlapping user samples but little overlap in user features. Specifically, VFL involves vertically partitioning datasets by feature dimension, extracting data segments where users are the same but features differ for training. For example, a bank and an e-commerce platform operating in the same region may have a significant overlap in user groups, but the bank records financial behavior data, while the e-commerce platform records shopping behavior data. Through VFL, these different features can be combined without sharing raw data, thereby enhancing the model's capabilities.

The advantages of VFL lie in its ability to utilize diverse feature data to build more comprehensive models, particularly suitable for joint marketing, risk management, and personalized recommendation applications. Using secure multi-party computation (SMC) and homomorphic encryption techniques, VFL can perform joint model training without directly exchanging raw data, ensuring data privacy protection. However, the implementation of VFL is complex, requiring solutions for synchronized computation, encrypted communication, and robust privacy protection. Synchronized computation demands real-time coordination among parties, while encrypted communication requires efficient encryption and decryption algorithms to ensure data security during transmission.

For recommendation systems, VFL is particularly useful when different platforms with overlapping user bases want to enrich their feature sets. For instance, a collaboration between a social media platform and a video streaming service could leverage VFL to combine user interaction data from both platforms. This would enable the creation of highly personalized content recommendations that consider both social media interactions and video viewing habits, enhancing user engagement and satisfaction. Another example is in the automotive industry, where car manufacturers and insurance companies can combine driving behavior data and insurance claim records to develop personalized insurance plans and safety recommendations for drivers.

Transfer Federated Learning (TFL) combines the principles of transfer learning and federated learning, suitable for scenarios where there is minimal overlap in both user samples and user features. In such cases, transfer learning is used to address the lack of data or labels instead of partitioning the data. For example, an e-commerce company in China and a social media application in the United States may have minimal user overlap due to geographical constraints, and their data features may only partially overlap due to the different nature of their businesses. To conduct effective federated learning in this context, transfer learning is introduced to address the issue of small data volumes and insufficient labeled samples, thereby improving model effectiveness.

Transfer learning is particularly useful for optimizing task performance when related data is insufficient. For instance, a hospital's radiology department may struggle to collect enough X-ray scans to build a robust diagnostic system. In such cases, transfer learning can leverage related tasks, such as image recognition, to aid in developing the diagnostic system. Through federated transfer learning, data privacy is maintained while transferring knowledge from auxiliary tasks to the target task, addressing the issue of limited data.

In recommendation systems, TFL can be instrumental in scenarios where different domains or industries wish to share insights without directly sharing data. For example, an online education platform and a job recruitment site could use TFL to enhance their recommendation systems. The education platform could leverage insights from job placement data to recommend relevant courses to students, while the recruitment site could use educational performance data to improve job matching algorithms. This cross-domain knowledge transfer can significantly improve the quality of recommendations in both platforms. Additionally, TFL can be used in the travel industry where airlines and hotels can share booking data to offer personalized travel packages to customers, enhancing their overall travel experience.

Horizontal Federated Learning, Vertical Federated Learning, and Transfer Federated Learning offer flexible and efficient solutions tailored to different data distributions and application needs. These modes ensure data privacy while enabling collaborative data and model optimization across institutions and fields. By adopting these federated learning approaches, industries can fully utilize data resources without sharing raw data, enhancing model performance and data security. Federated learning not only addresses privacy and security concerns but also promotes collaboration and innovation across organizations and sectors, breaking down data silos.

1.1.3 Cross-Device and Cross-Silo Federation

In federated learning (FL), cross-device and cross-silo are two primary application scenarios. These scenarios address different data distribution and application needs, providing flexible and efficient solutions to achieve effective model training while ensuring data privacy.

Cross-device federated learning is primarily applied to large-scale device networks, such as smartphones and Internet of Things (IoT) devices. Thousands of devices distributed across various geographical locations generate data independently, which is valuable for training. The core idea of cross-device federated learning is to distribute the model training tasks to individual devices, allowing these devices to independently train the model on their local data, and then send the trained model parameters to a central server for aggregation. This approach leverages distributed computing resources while mitigating the privacy risks associated with transmitting data over the network.

The significant advantage of cross-device federated learning lies in its robust data privacy protection. Since the data never leaves the device, user privacy is fully preserved, avoiding the risks of data breaches associated with centralized data storage. Additionally, cross-device federated learning reduces the computational load on the central server by utilizing edge computing resources, enhancing the overall system efficiency. Another advantage is its scalability, as it can support millions of devices participating in federated learning, thereby significantly expanding the coverage and diversity of training data. In practical applications, cross-device federated learning has been widely adopted in fields such as smartphones and smart homes. For instance, Google's Gboard uses cross-device federated learning to continuously improve the predictive performance and user experience of the keyboard. Smart home devices, such as smart thermostats and smart bulbs, use cross-device federated learning to enhance their intelligence and user satisfaction.

Another notable hardware development supporting cross-device federated learning is Apple's M1 chip, which significantly boosts the capability of AI processing directly on the device. The M1 chip's Neural Engine is designed to accelerate machine learning tasks, enabling efficient on-device computation, which is crucial for federated learning. By leveraging such high-performance chips, devices can locally train models faster, reducing latency and improving the overall user experience. Additionally, this on-device processing ensures that personal data remains on the device, enhancing privacy protection by minimizing the need to transmit sensitive data to central servers.

In practical terms, the M1 chip allows more complex federated learning models to be trained on individual devices without sacrificing performance. For instance, in personalized recommendation systems, an M1-powered device can locally process a user's interaction data, such as browsing behavior or content preferences, to refine recommendation algorithms. The device then sends encrypted model updates, rather than raw data, to a central server for aggregation. This setup ensures that user data stays private while benefiting from the collective learning of other devices in the network.

Such hardware advancements not only enhance the efficiency of cross-device federated learning but also address some of the key limitations of earlier implementations. By enabling devices to perform more computation locally, the reliance on network bandwidth is reduced, making federated learning more scalable and adaptable to diverse environments, including regions with limited network infrastructure. This is particularly important for maintaining consistent performance across a wide range of devices, from high-end smartphones like those powered by M1 chips to lower-powered IoT devices, while still ensuring robust privacy protections.

In the context of recommendation systems, cross-device federated learning plays a crucial role. It allows recommendation algorithms to be trained on user data directly on their devices, preserving privacy while continuously improving the recommendation quality. For example, video streaming services can utilize cross-device federated learning to analyze viewing patterns and preferences locally on users' devices, aggregating model updates to refine recommendation engines without compromising user data privacy. This approach is particularly beneficial in personalized content delivery, where user preferences are highly sensitive and diverse.

Cross-silo federated learning is suitable for scenarios where data is highly valuable but isolated, such as enterprises, hospitals, and financial institutions. These organizations typically possess large amounts of high-value data, but due to privacy and security concerns, this data is challenging to share. The core idea of cross-silo federated learning is to use secure multi-party computation (SMC) and homomorphic encryption technologies to enable data collaboration among different organizations without disclosing any sensitive information. This way, different organizations can jointly train high-performance machine learning models without sharing raw data.

The advantage of cross-silo federated learning is its ability to achieve efficient data collaboration and model training while ensuring data privacy and security. By leveraging the high-quality data from each organization, cross-silo federated learning can improve the accuracy and robustness of the models. For example, in the financial sector, different banks and financial institutions can jointly build fraud detection models and credit scoring systems through cross-silo federated learning, enhancing the security and accuracy of financial services. In the healthcare sector, different hospitals can share patient treatment data and genetic data, collaboratively developing disease prediction models and personalized treatment plans through cross-silo federated learning, thereby improving the quality of healthcare services and patient health outcomes. In corporate collaboration, different companies can share market data and user behavior data through cross-silo federated learning to jointly optimize marketing strategies and product designs, enhancing market competitiveness.

Comparing cross-device and cross-silo federated learning, several key differences emerge. Cross-device federated learning typically involves a vast number of devices with relatively smaller and less complex data, focusing on tasks that benefit from a wide distribution of data points. In contrast, cross-silo federated learning deals with fewer but more complex and high-value datasets from organizational silos, requiring sophisticated privacy-preserving techniques like SMC and homomorphic encryption. Furthermore, cross-device federated learning is highly decentralized, leveraging edge computing power, while cross-silo federated learning often involves coordination among fewer, more powerful institutional servers.

For recommendation systems, cross-device federated learning is often more suitable due to its ability to handle the dynamic and dispersed nature of user data. By enabling on-device learning, it ensures that recommendation models are continuously updated with the latest user interactions, providing timely and relevant recommendations. This is particularly important in environments like e-commerce, social media, and personalized content services, where user preferences can change rapidly, and privacy concerns are paramount.

In contrast, cross-silo federated learning can be beneficial in situations where rich, multifaceted data from various domains needs to be integrated to improve recommendation quality. For example, combining financial data from banks with shopping data from e-commerce platforms through cross-silo federated learning can create comprehensive profiles that enhance credit scoring and personalized financial product recommendations. However, the complexity and cost of implementing such systems are higher due to the need for advanced encryption and synchronization mechanisms.

In brief, cross-device and cross-silo federated learning provide flexible and efficient data processing and model training solutions for large-scale distributed devices and high-value data silos, respectively. These two modes effectively address data privacy and security issues while breaking down data silos through collaborative learning, promoting cross-organizational and cross-domain cooperation and innovation. As technology continues to evolve and improve, cross-device and cross-silo federated learning are expected to play crucial roles in more fields, driving the development of the digital economy and societal progress. For artificial intelligence researchers and beginners, understanding and mastering these two federated learning modes are key to advancing related research and applications.

1.1.4 Challenges in Federated Learning

Federated learning (FL) represents an advanced machine learning paradigm that enables collaborative model training across devices or organizations without sharing raw data, thereby offering substantial potential for data privacy protection. Despite its promise, FL faces numerous practical challenges, including data heterogeneity, communication overhead, privacy and security, computational resources and energy consumption, and legal and compliance issues.

Data heterogeneity is one of the primary challenges in FL. In FL, the data distributions among different participants can vary significantly, which impacts the effectiveness of model training and the performance of the final model. Traditional machine learning assumes data is independently and identically distributed (IID), but in FL, data on different devices or organizations often exhibits considerable variation in feature distribution and label distribution. For example, in cross-device federated learning, users' varied usage habits and environmental factors lead to significant differences in data generated on their devices; in cross-silo federated learning, differences in data sources, collection methods, and business focuses among organizations result in heterogeneous data. This heterogeneity complicates the aggregation process and impedes capturing global data patterns effectively. Furthermore, in federated environments, there are a large number of edge devices and the data held by these devices may be non-IID (non-Independent and Identically Distributed) [41]. For instance, in a smart medical system, the electronic medical record data structures for different diseases vary, presenting a substantial challenge for training these non-IID datasets.

For recommendation systems, this heterogeneity is particularly pronounced. User preferences for content, products, or services can vary widely across regions, cultures, and individual behaviors, complicating the task of creating a universally accurate recommendation model. Addressing this requires sophisticated techniques to balance and integrate diverse data sources.

Communication overhead is another critical issue in FL. The frequent exchange and synchronization of model parameters in FL generate significant communication overhead, particularly when involving numerous widely distributed devices. In cross-device federated learning, each device must upload locally trained model parameters to a central server and download the aggregated global model parameters, consuming considerable bandwidth and time. In cross-silo federated learning, data transmission and model parameter exchanges between organizations require efficient communication mechanisms to ensure real-time and stable training processes. Reducing communication frequency and data transmission volume while maintaining communication efficiency remains a significant challenge in FL.

In the realm of recommendation systems, increased communication overhead can significantly impact performance. Real-time updates and low-latency responses are crucial, and frequent model synchronization may cause delays, affecting user experience. Strategies such as model compression, asynchronous updates, and efficient communication protocols are essential to reduce the communication burden while maintaining system responsiveness.

Privacy and security issues are pivotal challenges in FL. Although FL inherently protects data privacy by keeping data local, privacy and security concerns persist in practical applications. During the transmission of model parameters, analyzing gradient information might still infer parts of the original data. Techniques like differential privacy and homomorphic encryption have been integrated into FL to enhance privacy protection, but they also increase computational complexity and communication overhead. Moreover, FL systems are vulnerable to attacks from malicious participants who may upload incorrect model parameters to degrade the global model's performance or employ data poisoning attacks to manipulate training outcomes. Developing robust and attack-resistant FL systems that ensure privacy and security is a crucial research focus.

When dealing with recommendation systems, privacy and security become even more critical. These systems handle highly sensitive user data, including browsing history, purchase records, and personal preferences. Ensuring that user data remains private and secure throughout the federated learning process involves implementing advanced privacy-preserving techniques and robust security measures, which can introduce additional computational overhead and complexity.

Computational resources and energy consumption present additional challenges for FL. FL relies on local computational resources of participating devices or organizations for model training. For resource-constrained devices (such as smartphones and IoT devices), performing complex model training tasks can impose a substantial computational burden and increase energy consumption. In cross-device federated learning, device heterogeneity means that some devices are limited in computational power, storage capacity, and battery life, affecting FL's efficiency and effectiveness. In cross-silo federated learning, disparities in computational resources and infrastructure among organizations must be coordinated and optimized to achieve efficient FL.

For recommendation systems, the demand for frequent model updates and the processing of large volumes of data can strain the resources of edge devices. Efficient algorithms that optimize computational load and energy consumption are necessary to make federated learning feasible on resource-constrained devices.

Legal and compliance issues also pose significant challenges for FL. During cross-organization and cross-region collaboration, FL must navigate diverse legal and regulatory frameworks across different countries and regions. Particularly in data privacy and security, varying laws such as the European Union's General Data Protection Regulation (GDPR) [18] impose strict requirements on data collection, storage, and processing. FL applications must adhere strictly to these regulations to avoid legal and compliance risks. Additionally, in scenarios involving sensitive data (such as medical and financial data), FL must consider data ethics and obtain user consent to ensure transparency and legality in data usage.

For recommendation systems, navigating legal and compliance requirements is equally critical, especially when handling user data across different jurisdictions. Adhering to various data protection laws and regulations while ensuring seamless operation requires careful planning and implementation of compliance strategies. Ensuring user consent and transparency in data usage is essential to maintain trust and meet legal obligations.

To summarize, while FL holds great promise in data privacy protection, model performance enhancement, and cross-organization collaboration, it faces substantial practical challenges. Addressing data heterogeneity, communication overhead, privacy and security, computational resources and energy consumption, and legal and compliance issues requires in-depth research and innovative solutions. For recommendation systems, these challenges are even more pronounced due to the need for real-time, personalized user interactions. Overcoming these obstacles will be critical for the successful deployment of federated learning in recommendation systems, ensuring both user satisfaction and data privacy protection.

1.2 Recommender System

When discussing cross-device federated recommendation, the concept of recommender systems (RecSys) is indispensable. For cross-device federated recommendation, the topic of recommender systems is unavoidable. In this section, we will provide a comprehensive overview of recommender systems, which are a crucial component of various online services such as e-commerce, social media, and content streaming platforms. Recommender systems aim to predict user preferences and suggest relevant items, thereby enhancing user experience and engagement.

The concept of recommender systems originated in the fields of information retrieval and data mining. Early methods included collaborative filtering, content-based filtering, and hybrid approaches. As technology has advanced, recommender systems have evolved, adopting more sophisticated algorithms and models to better meet user needs. In the subsequent sections, we will explore traditional recommendation methods, sequential recommendation, and session-based recommendation, analyzing their strengths, limitations, and application scenarios. Additionally, we will introduce graph neural network (GNN)-based recommendation methods, which leverage graph structures to model complex relationships between users and items, thereby improving recommendation accuracy. Finally, we will discuss the current challenges faced by recommender systems, including scalability, data sparsity, privacy concerns, and real-time processing requirements. By understanding these fundamental concepts and challenges, readers will gain a comprehensive understanding of the principles of recommender systems and the common methods used in the field.

1.2.1 Background and Concepts

The inception of recommender systems can be traced back to the rapid proliferation of the Internet and online services. With the exponential growth in the number of Internet users, the volume of data generated and accessed by users online has also exploded. This phenomenon of information overload made it difficult for users to find content they were interested in, thus diminishing user experience and platform engagement. In response to this challenge, recommender systems emerged, aiming to analyze user behavior and preferences to provide personalized content recommendations, thereby enhancing user experience.

The earliest recommender systems appeared in the 1990s when some e-commerce and content platforms began using collaborative filtering techniques to recommend products and content to users. In 1994, the GroupLens research group at the University of Minnesota launched the first automated recommender system, GroupLens [29]. This system introduced collaborative filtering as a key technology for recommender systems, marking the inception of automated collaborative filtering.

In 1997, Resnick et al. [30] first introduced the term "recommender system" (RecSys), which has since been widely adopted and has become a significant area of research.

In 1998, Amazon.com deployed an item-based collaborative filtering algorithm, scaling recommender systems to serve millions of users and handle millions of products while producing high-quality recommendations. Amazon's system analyzed the similarities between items to recommend products users might be interested in, significantly boosting purchase rates and platform sales.

In 2005, Adomavicius et al. [1] published a comprehensive review, categorizing recommender systems into three main types: content-based recommendations, collaborative filtering recommendations, and hybrid methods. This classification provided a systematic framework for recommender system research, aiding researchers in understanding and developing recommendation technologies.

In 2016, YouTube published a paper [3] applying deep neural networks to recommender systems, enabling the identification of the most likely recommendation results from a vast pool of potential content. This advancement marked the entry of recommender systems into the era of deep learning, further enhancing recommendation accuracy and user satisfaction through more complex models and algorithms.

The application scope of recommender systems has continuously expanded, from product recommendations on e-commerce platforms to friend recommendations on social media and content recommendations on streaming platforms. Recommender systems have become crucial tools for enhancing user experience and increasing platform revenue. Modern recommender systems utilize not only traditional collaborative filtering and content-based filtering methods but also advanced techniques such as matrix factorization (MF), deep learning, and graph neural networks to improve accuracy and robustness. These technologies enable recommender systems to handle large-scale data and capture more complex relationships between users and items.

A recommender system is an information filtering technology designed to analyze user behavior data, preferences, and history to provide personalized content recommendations. These systems use various algorithms and techniques to predict a user's interest in certain items (such as products, movies, and articles), delivering the most relevant and valuable recommendations. The goal of a recommender system is to enhance user experience and engagement, thereby increasing platform loyalty and revenue. Below, we present the formal definition of the general recommendation model. The general recommendation model $f(\theta)$ is formulated to learn from historical user-item interactions, represented by the triplets (u_i, v_j, r_{ij}) where u_i denotes a user, u_j denotes an item, and u_{ij} signifies the user's rating or interaction with the item. The model's primary objective is to accurately model user preferences by analyzing these interactions and subsequently predicting a ranked list of items from the set $V \backslash V_{u_i}$, which comprises items that the user u_i has not yet engaged with. The model seeks to minimize the deviation $\delta(\cdot)$ between the predicted rating $f(u_i, v_j|\theta)$ and the actual observed rating r_{ij}. Formally, this objective is achieved by identifying the optimal model parameters θ^* that minimize the cumulative deviation across all user-item pairs in the training data:

$$\theta^* = \arg\min_{\theta} \sum_{u_i \in U, v_j \in V_{u_i}} \delta\left(f(u_i, v_j|\theta), r_{ij}\right) \tag{1.1}$$

This optimization process is critical for ensuring that the model can generalize effectively to novel items, thereby producing rankings that are well-aligned with the inferred preferences of the user. To facilitate comprehension, we outline the typical pipeline of a recommender system. The typical pipeline of a recommender system includes the following key stages:

- **Matching** This stage generates hundreds of candidate items from an extremely large item pool (millions or even billions), significantly reducing the scale. Considering the large scale of data input in this stage, and due to the strict latency restrictions of online serving, complicated algorithms cannot be adopted, such as very deep neural networks [3, 14]. In other words, models in this stage are usually concise. That is, the core task of this stage is to retrieve potentially relevant items with high efficiency and attain coarse-grained modeling of user interests. It is worth noting that a recommender system in the real world usually

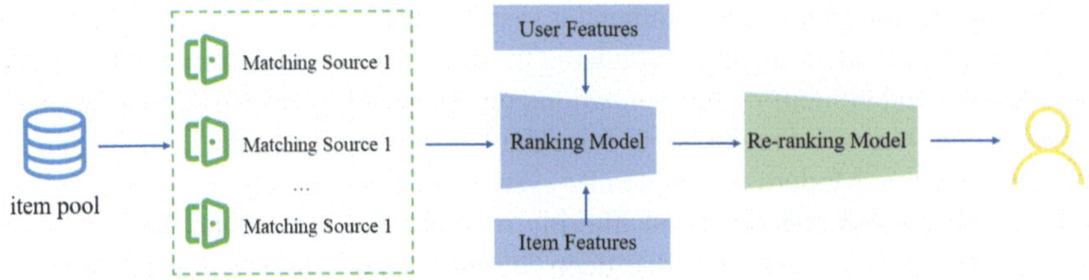

Fig. 1.3 The typical pipeline of recommender systems

contains multiple matching channels with multiple models, such as embedding matching, geographical matching, popularity matching, and social matching [9].

- **Ranking** After the matching stage, candidate items from different channels are merged into a single list and then scored by a ranking model. The ranking model sorts these items based on their scores, selecting the top few dozen. Given the relatively small number of items at this stage, the system can afford to use more complex algorithms to achieve higher recommendation accuracy. For example, rich features including user profiles and item attributes can be taken into consideration, and advanced techniques such as self-attention [13] can be utilized. The key challenge is to design appropriate models that capture complex feature interactions.
- **Re-ranking** Although the obtained item list after the ranking stage is optimized with respect to relevance, it may not meet other important requirements, such as freshness, diversity, and fairness [27]. Therefore, a re-ranking stage is necessary, typically removing certain items or changing the order of the list to meet additional criteria and business needs. The main focus at this stage is to consider multiple relationships among the top-scored items. For example, displaying similar or substitutable items closely together can lead to information redundancy (Fig. 1.3).

In the following sections, we will introduce several common types of recommender systems and discuss the current challenges in recommender system.

1.2.2 Traditional, Sequential, and Session-Based Recommendation

Recommender system (RecSys) assists users in selecting favorite items from massive data by considering varying user preferences. In accordance with the ability to discover short- or long-term preferences, RecSys can be categorized into Traditional RecSys, Sequence-based RecSys, and Session-based RecSys.

Traditional RecSys leverages long-term preferences through static user-item interactions, operating under the assumption that users maintain consistent interests over extended periods. These systems typically utilize collaborative filtering, content-based filtering, or hybrid methods to analyze user behavior and generate recommendations. Collaborative filtering can be further divided into user-based and item-based methods. User-based collaborative filtering identifies users with similar preferences and recommends items liked by those users, while item-based collaborative filtering recommends items similar to those the user has liked in the past. Content-based filtering, on the other hand, recommends items that are similar to those the user has previously expressed interest in, based on item attributes. Traditional RecSys excels in environments where user preferences do not change rapidly, such as recommending books, movies, or products in e-commerce.

In practical applications, traditional recommendation systems are widely used in e-commerce platforms like Amazon, which recommend products based on the purchasing behaviors of similar users, and in streaming services like Netflix, which suggest movies or TV shows similar to those previously watched by the user.

Conversely, Sequence-based RecSys and Session-based RecSys are designed to capture dynamic short-term preferences. Sequence-based RecSys utilizes the sequential dependency to model ordered historical interaction data, making it adept at understanding the evolving preferences of users over time. This approach is particularly useful in scenarios where user behavior follows a natural sequence, such as music streaming services, where users' listening patterns are sequentially dependent. Techniques such as recurrent neural networks (RNNs) and long short-term memory (LSTM) networks are often employed in Sequence-based RecSys to effectively model these sequential patterns and provide timely and relevant recommendations. A prime example of sequence-based recommendation can be seen in online learning platforms like Coursera, where the sequence of courses a user has taken can inform recommendations for future courses.

Session-based RecSys, on the other hand, explores co-occurrence dependencies to model both ordered and unordered data within clear session boundaries. This method focuses on capturing the immediate context of user interactions, making it highly effective for applications where user behavior is session-specific, such as online shopping sessions or video-on-demand platforms. For example, in an online shopping session, a user's interaction with various products can provide valuable insights into their immediate purchasing intent, allowing the system to recommend complementary or related products effectively. Below, we provide the definition of sequential recommendation. The model $f(\theta)$ learns interactions in chronological order $[v_{u_i,1}, \ldots, v_{u_i,j}]$ to capture user interests. It aims to determine the next sequence (or item) $\hat{V}_{u_i}^*$ that maximizes the score for u_i:

$$\forall u_i \in U, \; \hat{V}_{u_i}^* = \arg\max f(u_i, [v_{u_i,1}, \ldots, v_{u_i,j}] \| \theta) \tag{1.2}$$

In practice, session-based recommendation systems are essential in platforms like eBay, where they suggest additional items based on products viewed during the current shopping session, and in video streaming services like YouTube, which recommend videos related to the ones recently watched within the same browsing session.

Wang et al. [33] introduce the research progress of Sequence-based RecSys, tracing its evolution from traditional Markov chain models to popular neural network-based approaches. They provide concrete definitions and detailed comparisons of Sequence-based and Session-based RecSys in [34], highlighting the unique advantages and limitations of each method. Sequence-based RecSys are particularly strong in capturing temporal dynamics and order dependencies, whereas Session-based RecSys excel at modeling short-term interests within specific interaction sessions.

Apart from the above classification, various surveys review the research progress of RecSys from different perspectives. Zhang et al. [42] categorize deep learning-based RecSys by neural building blocks and deep hybrid models, offering a comprehensive introduction to the majority of model architectures. Given that user-item interactions can be naturally represented as a bipartite graph, numerous studies have leveraged graph neural networks (GNNs) to explore potential high-order connectivity between users and items. GNNs allow for the modeling of complex relationships and dependencies in the interaction data, facilitating the discovery of latent user-item connections that traditional methods might miss. Wu et al. [36] systematically present the development of GNN-based RecSys, providing a sophisticated categorization based on the inclusion of heterogeneous relations (e.g., knowledge graphs and social networks). These advanced methods enable the integration of rich contextual information, enhancing the accuracy and relevance of recommendations.

Moreover, point-of-interest (POI) recommendations guide users to latent favorite locations and can significantly increase foot traffic in business areas or tourist attractions. POI RecSys are particularly challenging due to the spatial-temporal dynamics involved and the need to consider users' contextual information such as location and time. Islam et al. [12] categorize POI-based RecSys in the context of deep learning and provide an extensive overview of popular POI datasets. They highlight the importance of integrating geographic information systems (GIS) and temporal patterns to improve the precision of location-based recommendations.

Understanding the different types of RecSys—Traditional, Sequence-based, and Session-based—and their respective methodologies is crucial for developing effective recommendation systems tailored to specific user behaviors and application contexts. These systems continue to evolve with advancements in machine learning and data representation techniques, offering increasingly sophisticated and accurate recommendations. By leveraging the strengths of each approach, it is possible to create hybrid systems that combine long-term user profiles with dynamic short-term preferences, ensuring a comprehensive and personalized user experience.

1.2.3 Graph Neural Network-Based Recommendation

In the current information society, recommender systems have become a crucial component of Internet services. They enhance user experience by recommending potentially interesting items to users and increase user engagement with the platform, thus driving business value. The core task of recommender systems is to provide personalized content based on users' historical behavior and preferences. Recently, the rapid development of deep learning, particularly the emergence of graph neural networks (GNNs), has significantly improved the performance of recommender systems. This is because much of the data in recommender systems naturally exhibits graph-structured characteristics, such as user-item interactions, social relationships among users, and semantic relationships among items. In these scenarios, traditional recommendation methods struggle to capture the complex relationships in the data, whereas GNNs, with their graph-based data representation and efficient graph computation models, can deeply explore these relationships, thereby enhancing recommendation accuracy and personalization.

GNNs are a class of neural network models designed to handle graph-structured data. Their core idea is to iteratively update node representations by aggregating information from neighboring nodes. In recommender systems, GNNs are typically applied through the following steps. First, the data in the recommender system is constructed into a graph structure, such as a user-item interaction graph, social network graph, or knowledge graph. Then, GNNs iteratively aggregate information from neighboring nodes through multiple propagation layers to update each node's embedding. This information aggregation process effectively captures high-order relationships between neighbors, resulting in richer and more accurate node representations. Through this method, GNNs excel not only in modeling user-item interactions but also in handling users' social relationships and semantic relationships among items, thereby improving the overall performance of recommender systems.

GNN-based recommender systems can be classified based on their application scenarios and the types of information utilized. The primary categories include user-item collaborative filtering, sequential recommendation, social recommendation, and knowledge graph-based recommendation. In user-item collaborative filtering, GNNs propagate information on user-item interaction graphs to capture high-order collaborative signals between users and items, thus enhancing recommendation accuracy. For example, GNNs can transform users' historical interaction behavior into high-quality user and item embeddings through multiple layers of propagation and aggregation. In sequential recommendation, GNNs construct a graph from users' behavior sequences to capture the sequential patterns between items, thereby predicting users' next preferences. For instance, users' browsing and purchasing behaviors on e-commerce platforms can be modeled as directed graphs, with GNNs capturing the temporal dynamics of user behavior. In social recommendation, GNNs leverage users' social relationships to model the diffusion of social influence among users, thereby improving user representations [7]. For example, interaction data from social media can be constructed as social graphs, with GNNs propagating social influence to enhance recommendation personalization. In knowledge graph-based recommendation, GNNs represent semantic relationships among items as knowledge graphs, utilizing their ability to handle heterogeneous graph structures to enhance item representations. For instance, book recommendation systems can use thematic relationships among books to construct knowledge graphs, with GNNs learning semantic embeddings to improve the relevance and diversity of recommendation results.

Recent years have seen the emergence of several seminal works in GNN-based recommender systems. Ying et al. [40] propose a graph convolutional network algorithm based on random walks, which was developed and deployed by Pinterest. It operates on a graph with 3 billion nodes and 18 billion edges, significantly improving user engagement. PinSage generates local subgraphs for nodes through random walks and aggregates node information from these subgraphs using graph convolutional networks, producing high-quality node embeddings. This successful application of GNNs in handling large-scale graph data demonstrates their powerful capabilities. Another important work is GC-SAN, which enhances sequential recommendation by stacking multiple self-attention layers on top of the item embeddings generated by GNNs. Xu et al. [38] first obtain initial item embeddings through GNNs and then applies self-attention mechanisms to further process these embeddings, capturing complex sequential patterns. This method significantly improves sequential recommendation performance. Wu et al. [35] propose a recommendation model that integrates user-item and social graphs. It learns user representations separately from the user-item interaction graph and the social graph and then combines these representations to enhance recommendation performance. This model effectively simulates the diffusion of social influence among users in social recommendation tasks.

GNN-based recommender systems exhibit strong representation learning capabilities, significantly enhancing recommendation performance. However, current methods still face some limitations, such as graph structure construction, neighbor aggregation strategies, and computational efficiency on large-scale graphs. Future research directions include more efficient graph structure construction methods, more complex neighbor aggregation strategies, and scalable computation on large-scale graphs. Continuous research and optimization will further advance the application of GNNs in recommender systems, enhancing their intelligence and accuracy. This will not only meet the growing demand for personalized user experiences but also drive overall progress in recommendation system technology.

1.2.4 Challenges in Recommender System

Despite significant advancements in the development of recommender systems, numerous challenges persist. These challenges encompass algorithmic difficulties, data issues, user privacy concerns, and system implementation complexities. One major challenge is the effective learning of user and item representations from massive datasets. As the number of users and items continues to grow, the data scale increases exponentially, making it difficult for traditional recommendation algorithms to handle such large-scale data. This challenge not only involves increased computational complexity but also efficiency in storage and retrieval. Researchers have proposed various methods based on distributed computing and large-scale parallel processing to address these issues. However, ensuring algorithmic accuracy and efficiency while processing massive datasets remains an unresolved problem.

Another critical challenge is the issue of data sparsity. In most recommender systems, users interact with only a small fraction of items, resulting in a highly sparse user-item matrix. This data sparsity severely affects the performance of recommendation algorithms, as they struggle to learn effective user and item representations from limited interaction data. To mitigate this problem, researchers have proposed methods to enhance data by leveraging social relationships, content information of items, and knowledge graphs to enrich user and item representations. However, these methods still face challenges in practical applications, such as the complexity of data integration and increased computational overhead.

Recommender systems also face the cold start problem, which refers to the difficulty in making recommendations for new users and new items due to a lack of sufficient historical interaction data. Common solutions include using social information of users, content information of new items, and similarity in interaction patterns to make recommendations. Additionally, researchers have proposed transfer learning techniques to transfer knowledge from other domains to the current domain to alleviate the cold start problem. However, efficiently implementing these methods without compromising recommendation accuracy remains a challenge.

Privacy and security issues are also significant challenges in recommender systems. User behavior data and personal information are crucial for recommendation algorithms, but their collection and use pose risks of privacy breaches. To protect user privacy, researchers have proposed various methods such as differential privacy and federated learning, which aim to achieve efficient recommendations while ensuring data privacy. Nevertheless, these methods face practical challenges like high computational complexity and significant communication overhead. Balancing user privacy protection and recommendation system performance remains a key research direction.

In practice, the adoption of privacy-preserving technologies in AI-enabled devices, such as smartphones and smart home systems, has gained significant momentum. Devices like AI-powered smartphones (e.g., Apple iPhones, and Google Pixel) and smart home assistants (e.g., Amazon Echo and Google Nest) process substantial amounts of user data, which is essential for personalized recommendations. Federated learning is one solution that enables these devices to train models locally on user data, enhancing privacy by only sharing aggregated model updates. However, this approach introduces several challenges, particularly in terms of computational efficiency and power consumption. While high-end devices like Apple's M1-powered smartphones can efficiently handle these tasks due to their advanced AI capabilities, lower-powered IoT devices may struggle with the computational demands. This imbalance in processing power across devices introduces limitations in cross-device environments, where not all devices can equally contribute to federated learning tasks.

Moreover, while techniques such as differential privacy are employed to enhance security, they often come at the cost of recommendation accuracy. The introduction of noise to obfuscate individual data points can diminish the overall effectiveness of the recommendation models, especially in cases where high precision is required. Current methods, though promising, still face significant obstacles in achieving the desired balance between protecting user privacy and delivering accurate, personalized recommendations.

Fairness and explainability are also important issues in recommender systems. Algorithms may produce biased recommendations due to data biases or flawed model designs, resulting in unfair treatment of certain user groups or neglecting some items due to insufficient interaction data. Additionally, the black-box nature of recommender systems makes it difficult to explain recommendation results, which not only affects user trust but also limits system tuning and improvement. To address these issues, researchers have proposed fair recommendation algorithms and explainable recommendation models that introduce fairness constraints and explainability mechanisms into the recommendation process to enhance system fairness and transparency.

1.3 Federated Recommender System

In the preceding sections, we have introduced the concepts and methodologies of federated learning (FL) and recommender systems (RecSys). Federated learning, a decentralized approach to machine learning, emphasizes data privacy by allowing model training to occur locally on user devices. Recommender systems, on the other hand, help users discover items of interest by analyzing user behavior and preferences. In this section, we bring these two domains together to explore Federated Recommender Systems (FedRec).

Federated Recommender Systems combine the strengths of FL and RecSys, providing personalized recommendation services while ensuring user privacy. This approach addresses the significant challenges of traditional recommender systems that rely on centralized data storage, which can lead to privacy concerns and data management issues. By leveraging federated learning techniques, FedRec can maintain data on local devices, exchanging only intermediate model parameters, thus protecting user privacy without compromising the quality of recommendations.

This section delves into the background and fundamental concepts of Federated Recommender Systems, examining the various communication architectures such as client-server and peer-to-peer (P2P) models. We will discuss the key technologies involved in implementing FedRec, including model updating and aggregation, privacy protection mechanisms, and communication efficiency optimization. Furthermore, we will explore the primary variants of FedRec—Horizontal, Vertical, and Transfer Federated Recommender Systems—highlighting their definitions, application scenarios, and respective advantages and challenges. Finally, we will compare FedRec with Distributed Recommender Systems (DRecSys) and Privacy-preserving Recommender Systems (PRecSys), elucidating their similarities, differences, and the unique benefits FedRec brings to the table.

By the end of this section, readers will have a comprehensive understanding of Federated Recommender Systems, their significance in the modern data-driven world, and the potential they hold for enhancing user privacy and recommendation accuracy in various domains.

1.3.1 Background and Concepts

With the rapid development of Internet technology, personalized recommender systems have become increasingly important in various applications such as e-commerce, social media, and content platforms. However, traditional recommender systems typically rely on centralized storage and processing of user data, which poses significant data management challenges and raises serious privacy concerns. To address these issues, federated learning (FL) technology has emerged and gradually been introduced into the field of recommender systems, forming Federated Recommender Systems (FedRec).

Federated Recommender Systems are designed to provide recommendation services while preserving user privacy. Participants in FedRec collaborate to train recommendation models by exchanging intermediate parameters instead of sharing raw data, ensuring that data remains on local devices and maximizing privacy protection. Ideally, the performance of recommendation models trained in a FedRec should be close to that of models trained in a centralized data environment, which can be formalized as follows:

$$|V_{\text{FED}} - V_{\text{SUM}}| < \delta \tag{1.3}$$

where V_{FED} represents the performance of the recommendation model in the FedRec, V_{SUM} represents the performance in traditional centralized systems, and δ is a small positive number.

In FedRec, participants' data is stored locally, and intermediate parameters are communicated between the server and the participants. There are two primary communication architectures used in FedRec research: client-server architecture and peer-to-peer architecture. The client-server architecture is the most common, relying on a trusted central server to perform initialization and model aggregation tasks. In each training round, the server distributes the current global recommendation model to a selection of clients. These clients use the received model and their local data for training and then send the updated intermediate parameters (such as model parameters and gradients) back to the server for global aggregation. The client-server architecture requires a central server to aggregate the parameters uploaded by the clients; therefore, if the server experiences a single point of failure, the entire training process is severely affected. Additionally, the server might infer sensitive information from the intermediate parameters, posing privacy risks.

To address the single point of failure issue inherent in the client-server architecture, Hegeds et al. designed a peer-to-peer communication architecture that does not involve a central server in the communication process. In each communication round, each participant broadcasts the updated intermediate parameters to some randomly selected online peers in the network and aggregates the received parameters into their own global model. This approach mitigates the single point of failure and privacy issues associated with a central server. However, the aggregation process occurring at each client significantly increases the communication and computational overhead for the clients.

The implementation of Federated Recommender Systems relies on several key technologies, including model updating and aggregation, privacy protection mechanisms, and communication efficiency optimization. In FedRec, participants perform local model training and periodically send the updated model parameters to a central server or other nodes in a peer-to-peer network. The server or nodes then aggregate these parameters to generate a new global model. A common aggregation method is federated averaging (FedAvg), which updates the global model by calculating the weighted average of the parameters uploaded by the clients.

To protect user privacy in Federated Recommender Systems, researchers have proposed various privacy protection mechanisms, such as differential privacy (DP) and secure multi-party computation (SMC). Differential privacy adds noise to the parameters during exchange to ensure that individual user data remains statistically indistinguishable, thus protecting privacy. Secure multi-party computation uses encryption techniques to ensure data security during parameter exchange, preventing participants from inferring other participants' raw data from the intermediate parameters. The privacy protection mechanisms of Federated Recommender Systems will be discussed in subsequent chapters.

Federated Recommender Systems involve significant communication overhead. To improve communication efficiency, researchers have proposed various optimization strategies, such as model compression, parameter quantization, and communication frequency adjustment. These strategies aim to reduce the volume and frequency of communication data, thereby lowering the system's communication overhead and improving overall efficiency. Model compression reduces the number of model parameters, parameter quantization decreases the bit representation of each parameter, and communication frequency adjustment reduces the number of communication rounds.

Federated Recommender Systems are widely applied in multiple domains, including e-commerce, social media, and content platforms. For instance, in e-commerce, Federated Recommender Systems can provide personalized product recommendations while protecting user shopping privacy. In social media, the system can recommend content and friends of interest while preserving user social behavior privacy. Federated Recommender Systems offer efficient and accurate recommendation services while safeguarding user privacy. With continuous development and optimization, Federated Recommender Systems are expected to play a crucial role in future recommendation technologies, further enhancing user experience and system performance.

1.3.2 Horizontal, Vertical, and Transfer Federated Recommendation

In Federated Recommender Systems (FedRec), Horizontal Federated Recommender Systems (HFedRec), Vertical Federated Recommender Systems (VFedRec), and Transfer Federated Recommender Systems (TFedRec) are three important variants. Understanding these systems' definitions and application scenarios is essential for comprehensively grasping their characteristics and potential value.

Horizontal Federated Recommender System (HFedRec) refers to a scenario where different data owners (such as enterprises and devices) collaboratively build a recommendation model without overlapping data samples. Specifically, each participant holds data with the same type of features (such as user and item features), but the user and item sets do not overlap. In HFedRec, participants share intermediate model parameters rather than actual data to collaboratively train the recommendation model. This approach not only protects user privacy but also enhances the performance of the recommendation model. Figure 1.4a illustrates the term "horizontal" with "horizontal partition" in traditional relational databases. Tabular data are split into groups by rows, and rows still possess complete columns. When we place this concept into the scenario of FedRec, each row comprises a user and interacted items (i.e., personal preference data). If an end device (e.g., mobile phone, tablet, or laptop) solely stores historical preference data for the current user, this type of FedRec is named "cross-user HFedRec," [39] as shown in Figs. 1.4b and 1.5. Conversely, if multiple personal preference data are gathered and stored by a platform, such FedRec is termed "cross-organization (or cross-silo) HFedRec," [39] as shown in Fig. 1.4c.

In order to facilitate the reader's understanding, we give the formula definition of HFedRec. In HFedRec, given federated clients C, each client retains private data $\mathcal{S}_c = \{V_c, U_c, V_c^{\text{feat}}, U_c^{\text{feat}}, R_c\}$, where U_c is the user set, V_c is the item set interacted with by U_c, V_c^{feat} (U_c^{feat}) represents item (user) features, and R_c denotes local user-

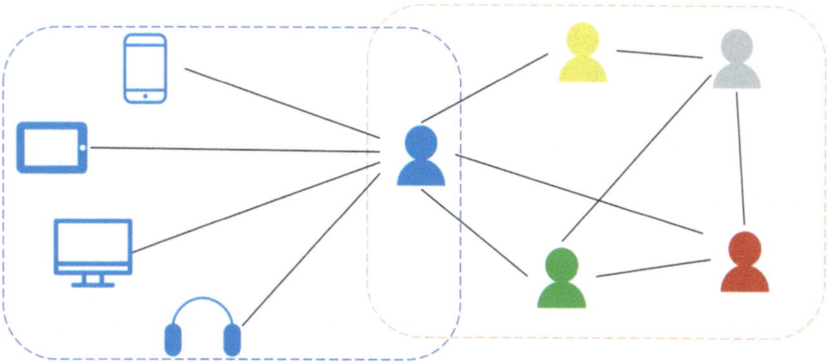

Fig. 1.4 The diagram of users and items node composition

item interactions (u_i, v_j, r_{ij}). For all $c, c' \in C$, where $c \neq c'$, HFedRec requires that $U_c \cap U_{c'} = \varnothing$ and that the feature categories are identical to ensure the same feature space. If the local number of users satisfies $|U_c| = 1$, it belongs to cross-user HFedRec; otherwise, it forms a cross-organization HFedRec system.

The primary application scenarios for Horizontal Federated Recommender Systems include cross-organization or cross-device data sharing, especially in contexts with a large amount of distributed user data, such as personalized recommendations in mobile applications. In these scenarios, HFedRec can improve recommendation model performance while protecting user privacy. For example, a mobile application can leverage historical data from different devices to provide personalized recommendations, thereby offering more accurate content. The advantages of HFedRec include effectively utilizing distributed data resources while avoiding the privacy risks associated with centralized data storage. However, HFedRec also faces challenges such as data imbalance among different clients and communication overhead. Additionally, since the data samples of each participant are different, effectively integrating these data to optimize the global model remains a critical issue.

Vertical Federated Recommender System (VFedRec) refers to a scenario where different data owners have overlapping user samples but different features, and they collaboratively build a recommendation model. Specifically, each participant holds part of the user or item feature data but may not have user-item interaction records. In this scenario, at least one client possesses interaction records, while other clients hold pre-aligned user or item features. This approach allows VFedRec to leverage the data resources of all participants, thereby improving the recommendation system's performance. Figure 1.5b illustrates the term "vertical" analogous to "vertical partition" in relational databases, where tabular data are split by columns. Based on this intuition, if organizations own overlapping user groups but different items, a vertical distribution emerges [37]. Note that before jointly training a VFedRec, organizations in Fig. 1.5b need to privately obtain the intersection of users [2]. Additionally, the federation where additional side information (e.g., user profiles and item attributes) is distributed across different clients also falls within the scope of VFedRec [8, 20].

In VFedRec, given federated clients C, each client retains private data $\mathcal{S}_c = \{V_c, U_c, V_c^{\text{feat}}, U_c^{\text{feat}}, R_c\}$, where not all variables are required. At least one client owns the interaction records R, while other clients may hold pre-aligned user profiles $\{U_c, U_c^{\text{feat}}\}$ or item attributes $\{V_c, V_c^{\text{feat}}\}$. As a result, VFedRec is applicable only in cross-organization federations.

The primary application scenarios for Vertical Federated Recommender Systems include cross-organization data cooperation, such as sharing user behavior data between different e-commerce platforms to enhance recommendation accuracy. For example, an e-commerce platform can use user feature data from its partners to improve its recommendation algorithm, thereby providing more personalized shopping suggestions. The advantages of VFedRec lie in fully utilizing the data resources of all parties, improving model performance, especially when there is strong synergy between different data holders. However, VFedRec also faces high

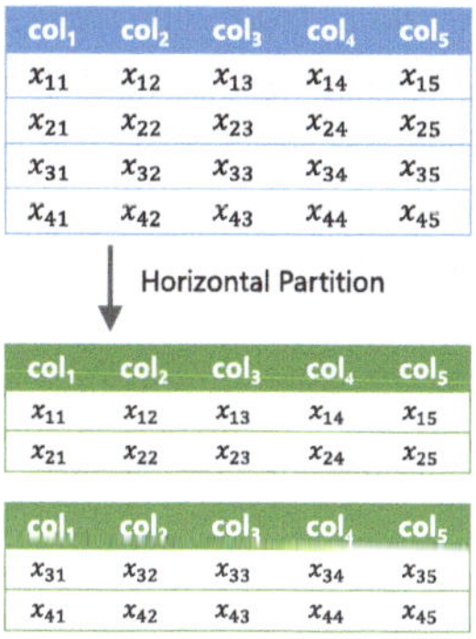

(a) Horizontal Partition of Data Table

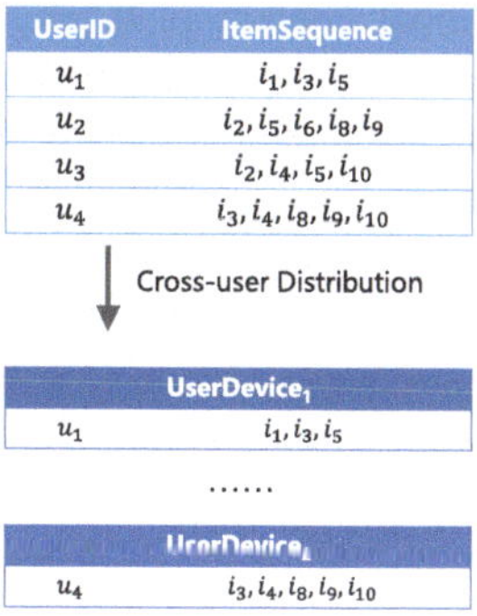

(b) Cross-user HFedRec

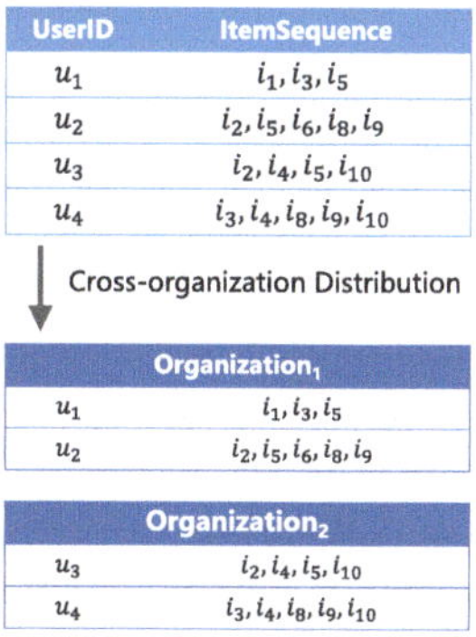

(c) Cross-organization HFedRec

Fig. 1.5 An analogy of horizontal data distribution between tabular and preference data

communication and computation overhead, particularly when frequent exchanges of intermediate parameters are required. Additionally, issues such as consistency and data alignment between different participants' data features need special attention.

Transfer Federated Recommender System (TFedRec) combines transfer learning and federated learning to address the issue of inconsistent data distribution. In this system, knowledge is learned from a source domain's data and applied to a target domain to improve the recommendation performance in the target domain. TFedRec enables knowledge transfer between the source and target domains, thereby enhancing the recommendation quality in the target domain, even when there are significant differences in data distribution.

Transfer Federated Recommender Systems are particularly suitable for scenarios with significant differences in data distribution, such as sharing user behavior data between different countries or regions. The advantage of TFedRec is that it can effectively utilize data from the source domain to improve the recommendation quality in the target domain, even when the target domain data is scarce. For example, a global content platform can use data from one region to improve recommendations in other regions. However, TFedRec also faces challenges such as model bias and adaptation issues during the transfer learning process. Researchers need to design reasonable transfer strategies and model adjustment methods to achieve effective knowledge transfer between different data distributions.

1.3.3 Distributed, Privacy-Preserving, and Federated Recommendation

In the field of recommender system research, Federated Recommender Systems (FedRec), Distributed Recommender Systems (DRecSys), and Privacy-preserving Recommender Systems (PRecSys) are three important and related concepts. Federated Recommender Systems have significant advantages in distributed computing and privacy protection, combining the computational capabilities of Distributed Recommender Systems with the security of Privacy-preserving Recommender Systems (Fig. 1.6).

Let us first discuss the relationship between FedRec and Distributed Recommender Systems (DRecSys). FedRec has a close relationship with DRecSys, which maintains a group of computing nodes and makes full use of each node's computing resources to learn users' preference collaboratively [4, 5]. We will list their similarities and differences in this section.

DRecSys and FedRec are similar in distributed computing and architecture: **(i)** A single computing node cannot meet the requirement of scalable storage capacity on account of the increasing proportion of users and items. As datasets and tasks are decentralized among multiple computing nodes, both can effectively reduce computational overhead by parallel computation. **(ii)** Both are applicable to parameter server (PS) and peer-to-peer (P2P) architectures. PS architecture composed of a server node and client nodes can regulate globally shared model

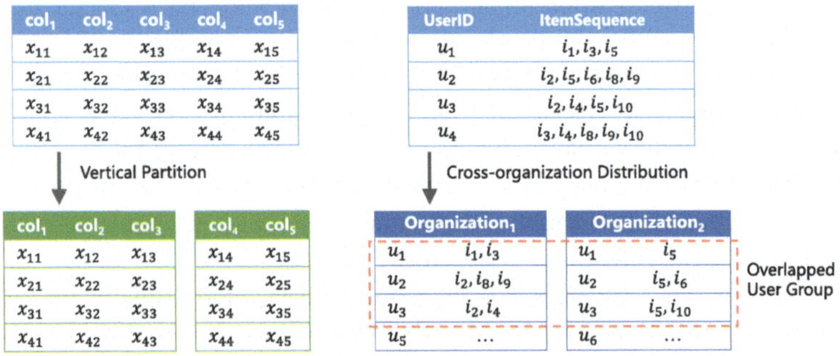

(a) Vertical Partition of Data Table (b) Cross-Organization VFedRec

Fig. 1.6 An analogy of vertical data distribution between tabular and preference data

parameters; P2P architecture composed of connected client nodes can reflect higher privacy security for the eliminated server node.

FedRec can be regarded as special DRecSys with the privacy constraint. Their differences consist in the following: **(i)** FedRec pays more attention to users' privacy. Most of DRecSys focus on improving distributed model accuracy and communication efficiency [11, 31], while FedRec gives priority to privacy protection. **(ii)** Clients/nodes play different roles. In DRecSys, clients generally refer to computing nodes in a computer cluster, and the server has absolute control over them. However, in FedRec, clients are not only computing nodes but data providers. They can regulate their own data independently and decide when to quit and join the federation. **(iii)** Communication overhead is more considerable in FedRec. On account of user devices participated in the cross-user federation, their network connections may be unstable and across long-distance areas. Oppositely, computing nodes in DRecSys are often close to each other and have a stable network environment. **(iv)** Raw data are unbalanced among clients in FedRec. Federated clients' data are locally generated rather than uniformly allocated by the server. Data amount and distribution characteristics may vary significantly, known as the non-IID phenomenon [19].

Does a Federated Recommender System belong to a Distributed Recommender System? The answer is "Yes." While maintaining the advantages of DRecSys, FedRec faces specific challenges. Except for the communication efficiency and non-IID phenomenon, FedRec may need to trade off between privacy and model performance when privacy-preserving techniques are applied.

Next, we will discuss another important similar concept—Privacy-preserving RecSys (PRecSys) and FedRec. We will give a brief overview and then point out the implications of FL for PRecSys.

According to different privacy-preserving techniques, traditional PRecSys can be divided into the following: **(i)** Cryptography-based PRecSys. Erkin et al. [6] utilize data packing and a semi-trusted third party to develop a cryptographic protocol. Preference data are uploaded under encryption to the recommender server. Kikuchi et al. [16] exploit a quasi-homomorphic mechanism to transmit ciphertexts of user similarities between two data parties. Studies on privacy-preserving matrix factorization [17, 24, 25] explore Yao's Garbled Circuits, Oblivious Transfer (OT), and homomorphic encryption to encrypt item-rating pairs from users. **(ii)** Obfuscation-based PRecSys. Polat et al. [28] allow the server to collect private preference data from users with a randomized perturbation scheme. McSherry et al. [23] apply centralized differential privacy to RecSys with more rigorous theoretical guarantee. Kikuchi et al. [15] utilize Randomized Response (RR) to disguise local preference data before submitted to the server. The authors devise a posterior probability distribution function to reconstruct the original distribution from disguised data.

Though the above studies obfuscate or encrypt raw preference data, private data leave local devices and are exposed in public. McMahan et al. [21] propose FL, which provides a new idea for PRecSys. In FedRec, only intermediate results of model training are communicated between the server and clients, which effectively avoids privacy risks arising from centralized storage of user behavior data.

(a) Communication Scheme of Traditional PRecSys (b) Communication Scheme of FedRec

Fig. 1.7 Different communication schemes between PRecSys and FedRec

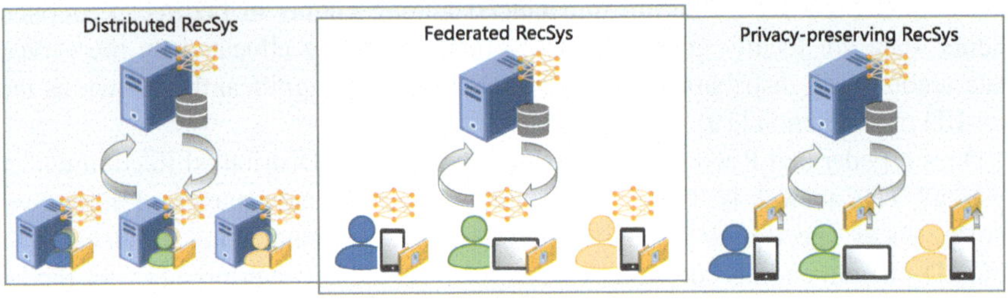

Fig. 1.8 Comparison of DRecSys, PRecSys, and FedRec

Therefore, for the cross-user distribution, the main difference between FedRec and PRecSys lies in the communication scheme, as shown in Fig. 1.7. Note that past cryptography- and obfuscation-based approaches are still necessary to ensure the security of FedRec.

Does a Federated Recommender System belong to a Privacy-preserving Recommender System? The answer is also "Yes." FedRec is a further improvement on traditional PRecSys. Moreover, we draw Fig. 1.8 to describe the relationship among FedRec, DRecSys, and PRecSys. As a dominant component of FedRec, HFedRec is promising to digest both advantages of DRecSys and PRecSys.

1.4 Summary

This chapter first introduces the basic concepts of federated learning (FL), emphasizing its significance in the era of big data and artificial intelligence. FL is a distributed machine learning method designed to address the challenges of traditional centralized data processing methods, such as data privacy issues, security risks, high data transmission costs, and processing delays. By allowing model training across multiple devices or organizations without sharing raw data, FL ensures that data remains local, sharing only model updates, thereby significantly reducing the risk of data breaches. FL includes three main collaborative modes: Horizontal Federated Learning (HFL), Vertical Federated Learning (VFL), and Transfer Federated Learning (TFL), each suitable for different data distribution and application scenarios.

Next, the chapter introduces the fundamental concepts of recommender systems. Recommender systems analyze user behavior and preferences to provide personalized content recommendations, enhancing user experience and engagement. Early methods of recommender systems include collaborative filtering, content-based filtering, and hybrid recommendation methods. As technology has advanced, recommender systems have increasingly adopted more sophisticated algorithms and models, such as sequential recommendation, session-based recommendation, and graph neural network-based recommendation methods, to better capture user needs and provide accurate recommendations.

Building on these two technologies, we introduce Federated Recommender Systems (FedRec), which combine the advantages of federated learning and recommender systems to provide efficient recommendation services while protecting user privacy. Federated Recommender Systems allow participants to collaboratively train recommendation models by exchanging intermediate parameters instead of raw data, ensuring that data remains local and user privacy is protected. The main communication architectures of Federated Recommender Systems include client-server and peer-to-peer architectures, each with its own advantages and disadvantages.

Despite the significant potential of Federated Recommender Systems, they still face numerous challenges in practical applications, such as data heterogeneity, communication overhead, privacy and security issues, computational resources, energy consumption, and legal and compliance issues. As Andrew Ng, a leading expert in AI and machine learning, stated, "Data is the new oil," but at the same time, "It needs to be refined with responsibility." This highlights the delicate balance between leveraging vast amounts of data for effective machine learning, like in recommendation systems, while ensuring that privacy is not compromised. Federated learning provides a mechanism to maintain this balance by keeping data decentralized and reducing the risk of data breaches.

Furthermore, in a discussion on federated learning, Ng emphasized, "The future of AI is increasingly on-device, where more computation happens locally, protecting data privacy while leveraging user interactions to improve AI models." This sentiment resonates strongly with the cross-device federated recommendation system context. Devices such as smartphones and IoT devices can train recommendation models locally, improving personalization while ensuring that sensitive user data remains on the device. However, as Ng and other experts have pointed out, "The challenge is not only technical but also legal and ethical," meaning that in addition to developing efficient federated models, addressing legal and compliance issues in data governance is paramount for real-world applications.

By introducing and analyzing federated learning, recommender systems, and Federated Recommender Systems, this chapter provides readers with a comprehensive understanding of these foundational concepts. In the next chapter, we will explore learning paradigms in cross-device federated recommendation.

References

1. Adomavicius G, Tuzhilin A (2005) Toward the next generation of recommender systems: A survey of the state-of-the-art and possible extensions. IEEE Trans Knowl Data Eng 17: 734–749
2. Chen H, Laine K, Rindal P (2017) Fast private set intersection from homomorphic encryption. In: Proceedings of the 2017 ACM SIGSAC Conference on Computer and Communications Security, CCS '17. ACM Press, New York, pp 1243–1255
3. Covington P, Adams J, Sargin E (2016) Deep neural networks for youtube recommendations. In: Proceedings of the 10th ACM Conference on Recommender Systems, RecSys '16. Association for Computing Machinery, New York, pp 191–198. https://doi.org/10.1145/2959100.2959190
4. Das AS, Datar M, Garg A, Rajaram S (2007) Google news personalization: scalable online collaborative filtering. In: Proceedings of the 16th International Conference on World Wide Web, WWW '07. ACM Press, New York, pp 271–280
5. Davidson J, Liebald B, Liu J, Nandy P, Vleet TV, et al (2010) The youtube video recommendation system. In: Proceedings of the Fourth ACM Conference on Recommender Systems, RecSys '10. ACM, New York, pp 293–296
6. Erkin Z, Veugen T, Toft T, Lagendijk RL (2012) Generating private recommendations efficiently using homomorphic encryption and data packing. IEEE Trans Inf Foren Secur 7:1053–1066
7. Fan W, Ma Y, Li Q, He Y, Zhao E, Tang J, Yin D (2019) Graph neural networks for social recommendation. https://arxiv.org/abs/1902.07243
8. Flanagan A, Oyomno W, Grigorievskiy A, Tan KE, Khan SA, Ammad-Ud-Din M (2021) Federated multi-view matrix factorization for personalized recommendations. In: Machine Learning and Knowledge Discovery in Databases. Springer International Publishing, Cham, pp 324–347
9. Gao C, Zheng Y, Li N, Li Y, Qin Y, Piao J, Quan Y, Chang J, Jin D, He X, Li Y (2023) A survey of graph neural networks for recommender systems: challenges, methods, and directions. ACM Trans. Recomm. Syst. 1:1–51. https://doi.org/10.1145/3568022
10. Gray W, Zheng HR (1986) General principles of civil law of the people's Republic of China. Am J Comparat Law 34:715–743. https://doi.org/10.2307/840330
11. Gupta V, Choudhary D, Tang P, Wei X, Wang X, Huang Y, Kejariwal A, Ramchandran K, Mahoney MW (2021) Training recommender systems at scale: communication-efficient model and data parallelism. In: Proceedings of the 27th ACM SIGKDD Conference on Knowledge Discovery & Data Mining, KDD '21. ACM, New York, pp 2928–2936
12. Islam MA, Mohammad MM, Sarathi Das SS, Ali ME (2022) A survey on deep learning based point-of-interest (poi) recommendations. Neurocomputing 472:306–325
13. Kang WC, McAuley J (2018) Self-attentive sequential recommendation. In: 2018 IEEE International Conference on Data Mining (ICDM), pp 197–206. https://doi.org/10.1109/ICDM.2018.00035
14. Kang WC, McAuley J (2019) Candidate generation with binary codes for large-scale top-n recommendation. In: Proceedings of the 28th ACM International Conference on Information and Knowledge Management, CIKM '19. Association for Computing Machinery, New York, pp 1523–1532. https://doi.org/10.1145/3357384.3357930

15. Kikuchi H, Mochizuki A (2012) Privacy-preserving collaborative filtering using randomized response. In: Proceedings of the 2012 Sixth International Conference on Innovative Mobile and Internet Services in Ubiquitous Computing, IMIS '12. IEEE Computer Society, Washington, pp 671–676
16. Kikuchi H, Aoki Y, Terada M, et al (2012) Accuracy of privacy-preserving collaborative filtering based on quasi-homomorphic similarity. In: 2012 9th International Conference on Ubiquitous Intelligence and Computing and 9th International Conference on Autonomic and Trusted Computing, pp 555–562
17. Kim S, Kim J, Koo D, Kim Y, Yoon H, et al (2016) Efficient privacy-preserving matrix factorization via fully homomorphic encryption: extended abstract. In: Proceedings of the 11th ACM on Asia Conference on Computer and Communications Security, ASIA CCS '16. ACM, New York, pp 617–628
18. Kuner C, Bygrave LA, Docksey C (2016) The GDPR: Understanding the new EU general data protection regulation. Eur Data Prot Law Rev 2:225–238. https://doi.org/10.21552/EDPL/2016/3/4. https://edpl.lexxion.eu/article/EDPL/2016/3/4
19. Ma X, Zhu J, Lin Z, Chen S, Qin Y (2022) A state-of-the-art survey on solving non-iid data in federated learning. Future Gener Computer Syst 135:244–258
20. Mai P, Pang Y (2024) Privacy-preserving multiview matrix factorization for recommender systems. IEEE Trans Artif Intell 5:267–277
21. McMahan B, Moore E, Ramage D, Hampson S, Arcas BAY (2017) Communication-efficient learning of deep networks from decentralized data. In: Proceedings of the 20th International Conference on Artificial Intelligence and Statistics, PMLR, pp 1273–1282
22. McMahan HB, Moore E, Ramage D, Hampson S, y Arcas BA (2023) Communication-efficient learning of deep networks from decentralized data. https://arxiv.org/abs/1602.05629
23. McSherry F, Mironov I (2009) Differentially private recommender systems: building privacy into the netflix prize contenders. In: Proceedings of the 15th ACM SIGKDD International Conference on Knowledge Discovery and Data Mining, KDD '09. ACM Press, New York, pp 627–636
24. Nayak K, Wang XS, Ioannidis S, Weinsberg U, Taft N, Shi E (2015) Graphsc: parallel secure computation made easy. In: 2015 IEEE Symposium on Security and Privacy (SP). IEEE Computer Society, Los Alamitos, pp 377–394
25. Nikolaenko V, Ioannidis S, Weinsberg U, Joye M, Taft N, Boneh D (2013) Privacy-preserving matrix factorization. In: Proceedings of the 2013 ACM SIGSAC Conference on Computer & Communications Security, CCS '13. ACM Press, New York, pp 801–812
26. Parasol M (2018) The impact of china's 2016 cyber security law on foreign technology firms, and on china's big data and smart city dreams. Comput Law & Secur Rev 67–98. https://doi.org/10.1016/j.clsr.2017.05.022. https://www.sciencedirect.com/science/article/pii/S0267364917300791
27. Pei C, Zhang Y, Zhang Y, Sun F, Lin X, Sun H, Wu J, Jiang P, Ge J, Ou W, Pei D (2019) Personalized re-ranking for recommendation. In: Proceedings of the 13th ACM Conference on Recommender Systems, RecSys '19. Association for Computing Machinery, New York, pp 3–11. https://doi.org/10.1145/3298689.3347000
28. Polat H, Du W (2003) Privacy-preserving collaborative filtering using randomized perturbation techniques. In: IEEE International Conference on Data Mining. IEEE Computer Society, Los Alamitos, pp 625–628
29. Resnick P, Iacovou N, Suchak M, Bergstrom P, Riedl J Grouplens: an open architecture for collaborative filtering of netnews. In: Proceedings of the 1994 ACM Conference on Computer Supported Cooperative Work, CSCW '94. Association for Computing Machinery, New York, pp 175–186 (1994). https://doi.org/10.1145/192844.192905
30. Resnick P, Varian HR (1997) Recommender systems. Commun ACM 40:56–58
31. Rong H, Wang Y, Zhou F, Zhai J, Wu H, Lan R, Li F, Zhang H, Yang Y, Guo Z, Wang D (2020) Distributed equivalent substitution training for large-scale recommender systems. In: Proceedings of the 43rd International ACM SIGIR Conference on Research and Development in Information Retrieval. ACM Press, New York, pp 911–920

32. Saha S, Ahmad T (2021) Federated transfer learning: concept and applications. https://arxiv.org/abs/2010.15561
33. Wang S, Hu L, Wang Y, Cao L, Sheng Q, Orgun M (2019) Sequential recommender systems: challenges, progress and prospects. In: Proceedings of the 28th International Joint Conference on Artificial Intelligence, pp 6332–6338
34. Wang S, Cao L, Wang Y, Sheng QZ, Orgun MA, Lian D (2021) A survey on session-based recommender systems. ACM Comput Survs 54(7):1–38
35. Wu L, Sun P, Fu Y, Hong R, Wang X, Wang M (2019) A neural influence diffusion model for social recommendation. In: Proceedings of the 42nd International ACM SIGIR Conference on Research and Development in Information Retrieval, pp 235–244
36. Wu S, Sun F, Zhang W, Xie X, Cui B (2022) Graph neural networks in recommender systems: a survey. ACM Comput Surv 55:1–37
37. Wu Z, Li Q, He B (2022) Practical vertical federated learning with unsupervised representation learning. IEEE Trans Big Data 10(6):864–878
38. Xu C, Zhao P, Liu Y, Sheng VS, Xu J, Zhuang F, Fang J, Zhou X (2019) Graph contextualized self-attention network for session-based recommendation. In: IJCAI
39. Yang Q, Liu Y, Chen T, Tong Y (2019) Federated machine learning: concept and applications. ACM Trans Intell Syst Technol 10:1–19
40. Ying R, He R, Chen K, Eksombatchai P, Hamilton WL, Leskovec J (2018) Graph convolutional neural networks for web-scale recommender systems. In: Proceedings of the 24th ACM SIGKDD International Conference on Knowledge Discovery & Data Mining, pp 974–983
41. Zhang C, Xie Y, Bai H, Yu B, Li W, Gao Y (2021) A survey on federated learning. Knowl-Based Syst 216:106775. https://doi.org/10.1016/j.knosys.2021.106775. https://www.sciencedirect.com/science/article/pii/S0950705121000381
42. Zhang S, Yao L, Sun A, Tay Y (2019) Deep learning based recommender system: a survey and new perspectives. ACM Comput Surv 52:1–38

Chapter 2
Learning Paradigms in Cross-Device Federated Recommendation

Abstract This chapter explores learning paradigms in cross-device federated recommendation systems, analyzing statistical machine learning, deep learning, meta-learning, and reinforcement learning (RL). It reviews key studies, highlighting models, techniques, and datasets used in FedRec. The chapter addresses challenges such as data heterogeneity, communication efficiency, and model adaptability. It emphasizes the importance of efficient aggregation strategies and advanced privacy-preserving techniques to handle diverse data and models. A comparative analysis of different approaches reveals their strengths and limitations, offering insights to advance federated recommendation systems. Finally, the chapter discusses future research directions to enhance system scalability and effectiveness.

2.1 Statistical Machine Learning

We put representative works from the model perspective in Table 2.1, including learning paradigm, recommendation models, recommendation problems, federated distribution types, and dataset types. We confirm dataset types by empirical studies in their own papers. Some publications are difficult to reflect their uniqueness in this table, such as FedRec$_{\mathrm{L}}$ [33], FedRec++ [32], PerFedRec [40], and PerFedRec++ [41]. More details can be seen in the following sections and Tables 2.3–4.1.

According to Table 2.1, matrix factorization is the dominant approach in machine learning-based HFedRec, aiming to decompose user embeddings $u_i \in \mathbb{R}^{d_u}$ and item embeddings $v_j \in \mathbb{R}^{d_v}$ from the sparse collaborative matrix $R \in \mathbb{R}^{N_U \times N_V}$. Matrix factorization (MF) is a foundational technique in recommender systems, designed to identify the latent factors that account for observed user-item interactions. The primary objective of MF is to approximate the large, sparse matrix R as the product of two lower-dimensional matrices, each representing users and items, respectively. This approximation is formalized as follows:

$$R \approx U \cdot V^T \tag{2.1}$$

X. Kong et al., *Cross-device Federated Recommendation*, Machine Learning: Foundations, Methodologies, and Applications,
https://doi.org/10.1007/978-981-96-3212-1_2

Table 2.1 Representative works in HFedRec from the model perspective

Publications	Type	Model	Prob	U/O	Datasets
FCF [2]	M	WMF	G1	U	Movie
JointRec [12]	D	CNN	G1	U	Movie
FedRec$_L$ [33]	M	PMF	G1	U	Movie
FedPOIRec [50]	D	CNN	S2	U	Check-in
FedRec++ [32]	M	PMF	G1	U	Movie
FedGME [49]	D	AE	G1	O	Stock
DFedRec [30]	M	PMF	G1	U	Movie
ReFRS [19]	D	AE	S1	U	Movie, Music
FedMF [5]	M	PMF	G1	U	Movie
CPF-POI [78]	D	MM	S2	U	Check-in
RFPG [10]	M	BPRMF	S2	U	Check-in
PrefFedPOI [87]	D, R	MM	S2	U	Check-in
DFSR [39]	M	PMF	G1	U	Movie
FedNewsRec [53]	D	MM	G1	U	News
LightFR [84]	M	MF-L2H	G1	U	Movie, Product
Efficient-FedRec [79]	D	MM	G1	U	News
FCF-BTS [23]	R	MAB	G1	U	Movie, Music, News
KG-FedTrans4Rec [71]	D	MM	S3	O	Movie, Music, Book
FPAE [37]	R	MAB	G1	O	Product
FedSeqRec [26]	D	MM	S1	O	Movie, Video
Fed2-UCB [60]	R	MAB	G1	U	Movie
Limited Negatives [47]	D	MLP	G1	O	Movie
FedPerGNN [72]	D	GNN	G3	U	Movie, Music
HPFL [73]	D	MLP	G1	O	Movie, Edu
PerFedRec [40]	D	GNN	G3	U	Movie, Book, Business
PFedRec [82]	D	MLP	G1	U	Music, Video
PerFedRec++ [41]	D	GNN	G3	U	Book, Check-in, Business
FedFast [45]	D	MLP	G1	U	Movie, Product, Business
GPFedRec [81]	D	GNN	G3	U	Movie, Music
FedMeta [7]	T	MAML	G1	U	Mobile Service
FedHGNN [75]	D	GNN	G3	U	Cite, Business
MetaMF [34]	T	MAML	G1	U	Movie, Product
DGREC [89]	D	GNN	G3	U	Movie, Book, Check-in
Fed4Rec [88]	T	MAML	G1	U	News

(continued)

Table 2.1 (continued)

Publications	Type	Model	Prob	U/O	Datasets
SemiDFEGL [54]	D	GNN	G2	U	Movie, Check-in, Business
FedIS [9]	T	Reptile	G1	U	Movie, Music, Check-in, Game
CdFed [86]	D	GNN	G2	U	Product
DP-PrivRec [70]	T	Reptile	G1	U	Movie, APP

Type: M, statistical machine learning; D, deep learning; T, meta-learning; R, reinforcement learning.
U/O: U, cross-user federation; O, cross-organization federation.
Model: PMF, probabilistic matrix factorization [58]; WMF, weighted MF [18].
BPRMF, Bayesian personalized ranking-based MF [55]; MAB, multi-armed bandits.
GNN, graph neural network; MLP, multilayer perceptron.
CNN, convolutional neural network; MM, mixture model; AE, autoencoder.
MAML, model-agnostic meta-learning [14].
Reptile, reptile [46]; MF-L2H, MF based on learning to hash [69].
Prob (recommendation problem): Referring to Table 2.2 for meanings of index codes.

Table 2.2 Index code of recommendation problems for Table 2.1

Category 1	Category 2	Index Code
General recommendation	Vanilla user-item interactions	G1
	User-item bipartite graph	G2
	Heterogeneous graph (e.g., social links between users)	G3
Sequential recommendation	Vanilla chronological order	S1
	Spatial-temporal POI	S2
	Knowledge graph	S3

where $U \in \mathbb{R}^{N_U \times d_u}$ represents the user matrix and $V \in \mathbb{R}^{N_V \times d_v}$ represents the item matrix. Each row in U corresponds to a user embedding, while each row in V corresponds to an item embedding. The goal of MF is to minimize the reconstruction error, typically quantified by the Frobenius norm:

$$\min_{U,V} \|R - U \cdot V^T\|_F^2 \tag{2.2}$$

In the context of HFedRec, matrix factorization must be adapted to a federated learning framework, where user data is distributed across multiple clients, necessitating a collaborative model training process without the centralization of raw data.

Weighted matrix factorization (WMF) extends the basic MF approach by incorporating varying confidence levels associated with different entries in the matrix R. This is achieved by introducing a confidence matrix C, which weighs the significance of each observed rating, thereby modifying the optimization objective as follows:

$$\min_{U,V} \sum_{(i,j)\in \text{observed}} C_{ij} \cdot \left(R_{ij} - u_i^T v_j\right)^2 + \lambda(\|U\|_F^2 + \|V\|_F^2) \tag{2.3}$$

where λ is a regularization parameter to mitigate overfitting.

Probabilistic matrix factorization (PMF) further enhances MF by framing it within a probabilistic model. PMF assumes a Gaussian distribution for the observed ratings and incorporates prior distributions over the latent factors for users and items. The corresponding objective function for PMF is derived from the log-likelihood of the observed data:

$$\log p(R \mid U, V) = -\frac{1}{2\sigma^2} \sum_{(i,j)\in \text{observed}} \left(R_{ij} - u_i^T v_j\right)^2 - \frac{\lambda}{2}\left(\|U\|_F^2 + \|V\|_F^2\right) \tag{2.4}$$

Under the parameter server architecture, the typical federated training steps of MF are listed as follows:

- **Step 1.** The server initializes global item embedding matrix $\boldsymbol{V} \in \mathbb{R}^{N_V \times d_v}$ and samples a batch of clients C to start the federated training process.
- **Step 2.** Selected clients C download latest $\boldsymbol{V}$ from the server and then utilize local implicit or explicit feedback to update $\boldsymbol{u}_i$ and compute $\nabla_{\boldsymbol{v}_j}$.
- **Step 3.** After clients complete local training in parallel, only item gradients are transmitted to the server.
- **Step 4.** The server aggregates gradients uploaded by clients and updates $\boldsymbol{V}$. Then the server distributes latest $\boldsymbol{V}$ to clients for the next communication round.
- **Step 5.** Repeat steps 2–4 until convergence.

Accordingly, involved data can be divided into the following: (i) Public parts mainly contain the global item embedding matrix $\boldsymbol{V}$ and gradients $\nabla_{\boldsymbol{v}}$ (client ID, round ID, and other auxiliary information [27] are omitted), which can be shared and transmitted between federated participants. (ii) Private parts are user embeddings and original user-item interaction data preserved in local devices. We will describe details from client and server sides in subsections.

2.1.1 Client-Side Training

Most of MF-based HFedRec are the federation of probabilistic matrix factorization (PMF) [58] or other variants [18, 62]. PMF formulates the loss function as follows:

$$L=\frac{1}{2}\sum_{i=1}^{N_U}\sum_{j=1}^{N_V} I_{ij}\left(r_{ij}-\boldsymbol{u}_i\boldsymbol{v}_j\right)^2+\frac{\lambda_U}{2}\sum_{i=1}^{N_U}\|\boldsymbol{u}_i\|_2^2+\frac{\lambda_V}{2}\sum_{j=1}^{N_V}\|\boldsymbol{v}_j\|_2^2 \tag{2.5}$$

where $I_{ij}=1$ when u_i rated v_j, and $I_{ij}=0$ otherwise. λ_U, λ_V are regularization parameters of U, V respectively. When updating $\boldsymbol{u}_i$ and $\boldsymbol{v}_j$ with gradient descent, we can derive the partial derivative:

$$\frac{\partial L}{\partial \boldsymbol{u}_i}=-\sum_{j=1}^{N_V} I_{ij}(r_{ij}-\boldsymbol{u}_i\boldsymbol{v}_j)\boldsymbol{v}_j+\lambda_U\boldsymbol{u}_i,\quad \frac{\partial L}{\partial \boldsymbol{v}_j}=-\sum_{i=1}^{N_U} I_{ij}(r_{ij}-\boldsymbol{u}_i\boldsymbol{v}_j)\boldsymbol{u}_i+\lambda_V\boldsymbol{v}_j \tag{2.6}$$

Owing to privacy-sensitive user embeddings required to be maintained locally [5], it is impossible to leverage all user embeddings to update item embeddings by $\boldsymbol{v}_j=\boldsymbol{v}_j-\eta\partial L/\partial\boldsymbol{v}_j$. For handling this issue, Ammad-ud-din et al. [2] first propose federated collaborative filtering (FCF) and define the basic training steps mentioned above. Only the current user embedding is utilized to calculate local item gradients. Most of studies [5, 11, 13, 30, 44, 77] follow the solution in FCF. Generally, $\nabla_{\boldsymbol{u}_i^t}$ and $\nabla_{\boldsymbol{v}_j^t}$ at t-th local epoch are computed like the following:

$$\nabla_{\boldsymbol{u}_i^t}=-\sum_{j\in V_{u_i}}\left(r_{ij}-\hat{r}_{ij}\right)\boldsymbol{v}_j^{t-1}+\lambda\boldsymbol{u}_i^{t-1},\quad \nabla_{\boldsymbol{v}_j^t}=-\left(r_{ij}-\hat{r}_{ij}\right)\boldsymbol{u}_i^{t-1}+\lambda\boldsymbol{v}_j^{t-1} \tag{2.7}$$

Especially, FedRec$_L$ [33] and FedRec++ [32] randomly sample unrated itemsF from V_u' and assign them virtual ratings. Instead of an entire item set, clients upload interacted item gradients with fake entries to the server. This innovative approach significantly enhances user privacy protection by obscuring actual user preferences, thus reducing the risk of privacy breaches during data transmission. Furthermore, it maintains the integrity and performance of the recommendation model by ensuring that the gradients submitted for aggregation are representative of the user's interactions, albeit with some noise introduced by the virtual ratings. This method also reduces the communication overhead as only selected gradients are transmitted, which is particularly beneficial in resource-constrained environments.

Yang et al. [76] explore the scenario that organization-partitioned preference data are heterogeneous, e.g., scoring ranges of $\{1, 1.5, \ldots, 4.5, 5\}$ and $\{1, 2, \ldots, 9, 10\}$. The local interaction data of two parties are treated as a target matrix and an auxiliary matrix, where the latter is served as complementary information to the former. This method effectively leverages the diverse data distributions to enhance the recommendation accuracy, accommodating different organizational contexts and user behavior patterns. By combining information from multiple sources, this approach can generate more robust and generalized models that perform well across varied datasets.

Zhang et al. [39] join a contrastive self-supervised task to the target recommendation task. The contrastive goal is to close the distance between positive embeddings of local users, social friends, local items, and global items. This integration of contrastive learning helps improve the robustness of the recommendation model by aligning similar user and item embeddings more closely, thereby enhancing the model's ability to capture relevant user preferences. The self-supervised learning component adds an extra layer of training that enriches the representation of data without requiring additional labeled data.

In a decentralized setting, Yang et al. [30] explore a decentralized setting where user-user collaboration depends on a social network constructed by a temporary server. According to social links, clients can exchange item gradients with their neighbors who have co-interacted items. This peer-to-peer approach facilitates more direct and efficient gradient exchange, leveraging existing social networks to improve the federated learning process's efficiency and effectiveness. By utilizing social network structures, DFedRec can enhance the accuracy of recommendations and reduce the reliance on a central server, thereby increasing the system's resilience to single points of failure.

The client-side training process in federated learning provides several key advantages. It ensures that sensitive user data remains local, significantly enhancing privacy and security. Local computation also allows for real-time updates and personalization, as models can be updated with the latest user interactions without waiting for centralized aggregation. This decentralized approach also reduces the risk of data breaches associated with centralized data storage and transmission, making the entire system more robust and trustworthy. Furthermore, it allows for scalable training across a vast number of devices, each contributing to the overall model's improvement while preserving individual data privacy.

2.1.2 *Server-Side Training*

At the r-th communication round, clients compute local item gradients in parallel and submit gradients to the server. Before updating the global item embedding matrix $\boldsymbol{V}$, the server needs to aggregate uploaded gradients:

$$\boldsymbol{v}_j^r = \boldsymbol{v}_j^{r-1} - \eta \nabla_{\boldsymbol{v}_j^r} = \boldsymbol{v}_j^{r-1} - \eta \sum_{i=1}^{N_C} \nabla_{\boldsymbol{v}_{ij}^r} \tag{2.8}$$

This aggregation process is crucial for maintaining the consistency and accuracy of the global model. By collecting and averaging the gradients from multiple clients, the server can effectively integrate the diverse local updates, ensuring that the global model benefits from the data distributed across all participating devices. This approach leverages the combined learning power of all clients, resulting in a more generalized and robust model.

For eliminating the precision loss caused by pseudo interacted items, Xu et al. in their work FedRec++ [32] selects some denoising clients to aggregate gradients $\nabla_{\boldsymbol{v}'_j}$ from U'_{v_j}. This selection process helps to mitigate the noise introduced by fake entries, enhancing the accuracy of the aggregated gradients and, consequently, the global model.

Differently, Yang et al. [30] select a temporary server to receive real and pseudo interacted items from other clients. Based on co-interacted items, the server constructs a social network containing user nodes and social links to guide decentralized collaboration, which is analogous to the global graph reconstruction in Sect. 2.2.3 with respect to GNN-based HFedRec. This decentralized approach reduces the dependency on a single central server, increasing the system's resilience and fault tolerance.

In addition, Li FCF et al. [2] and Chen A-RFRS et al. [6] analyze the Adam optimizer applied to MF. The server first aggregates received moments from clients, and then adaptive learning rate and global item embeddings are updated by aggregated moments. The use of advanced optimization techniques like Adam helps in accelerating the convergence of the model and improving its performance by adapting the learning rate dynamically based on the gradients' moments.

Server-side training provides several significant advantages. It centralizes the complex task of model aggregation, which simplifies the computation required at each client, making the process more efficient for devices with limited resources. This centralized approach also allows for more sophisticated and powerful computational resources to be utilized at the server, enabling the handling of larger models and more complex computations that would be infeasible on individual clients.

Moreover, the server can implement robust security measures to ensure the integrity and confidentiality of the aggregated data, enhancing the overall security of the federated learning process. By managing the aggregation centrally, the server can detect and mitigate potential security threats more effectively, such as identifying malicious clients attempting to disrupt the learning process.

In summary, server-side training in federated learning plays a critical role in aggregating and updating the global model based on the gradients submitted by clients. This process not only ensures the consistency and accuracy of the model but also leverages the combined computational power of all clients, resulting in a more robust and generalized model. The centralized aggregation at the server offers numerous advantages, including enhanced computational efficiency, improved security, and the ability to handle complex models and computations.

2.1.3 Summary and Challenges

Recent studies have explored federated MF with various privacy-preserving techniques, such as HE [5, 76], DP [44, 76], and SMC [13, 35], but paid little attention to MF itself. These privacy-preserving techniques are crucial in ensuring that sensitive user data remains protected during the federated learning process.

The first challenge is that **MF encounters a dilemma in fusing side information.** To address this issue, Yu et al. [10] privately incorporate geographical locations into MF to enhance POI recommendation performance. Effectively incorporating various types of side information, such as demographic data and temporal patterns, remains a complex task requiring further exploration.

Another significant challenge is that **sparse and large-scale user-item interactions account for the efficiency bottleneck**, especially in the resource-constrained federation. LightFR [84] is the first effort to highlight the factorization ability through learning to hash [69]. In this approach, users and items are represented as compact binary codes, allowing for efficient computation of user-item similarity in a discrete Hamming space. This method significantly reduces the computational load and enhances the scalability of federated MF.

In summary, while federated MF with privacy-preserving techniques offers significant advantages, it also presents various challenges that need to be addressed to optimize its performance and applicability. Continued innovation and collaboration are essential to fully realize the potential of federated MF in providing robust, scalable, and privacy-preserving recommendation systems.

2.2 Deep Learning

In contrast to statistical machine learning approaches, deep learning-based RecSys can effectively mine deeper user-item relationships with powerful nonlinear transformation ability [8]. The federated training steps of deep learning-based HFedRec can be summed up as follows:

- **Step 1.** The server initializes weights of the global model and samples a batch of clients to start the federated training steps.
- **Step 2.** Selected clients download weights of the global model from the server and then utilize local implicit or explicit feedback to update local models.
- **Step 3.** After clients complete local training in parallel, gradients or weights are transmitted to the server.
- **Step 4.** The server aggregates params uploaded by clients and updates the global model. Then the server distributes latest weights to clients for the next communication round.
- **Step 5.** Repeat steps 2 to 4 until convergence.

Apart from item embeddings, whether hidden layers' weights need aggregating also remains decision. The server can solely take charge of aggregating uploaded parameters without maintaining a global model. Then, we categorize HFedRec in terms of different types of neural building blocks.

2.2.1 Multilayer Perceptron Network

The multilayer perceptron network (MLP) is a concise neural network and basic module for other complex models [52]. It is also a commonly used and important module in federated recommendation systems. Therefore, we will first provide a brief introduction to the MLP.

Introduction Multilayer perceptron network (MLP) is a type of artificial neural network (ANN) and a fundamental model in deep learning. An MLP is a feed-forward neural network, meaning data flows through the network in one direction without any feedback loops. An MLP consists of at least three layers: an input layer, one or more hidden layers, and an output layer. Each layer comprises multiple neurons connected by weighted links. These connections enable the MLP to perform complex nonlinear transformations on input data, thus solving various classification and regression problems.

The basic concept of an MLP is inspired by biological neural networks. Biological neurons are connected via synapses, each with different conductive capacities, analogous to the weights in an MLP. When a neuron's received signals exceed a certain threshold, it activates, similar to how activation functions work in an MLP. The MLP adjusts these weights using training data to generate the desired output for given inputs.

In machine learning and data mining, MLPs are widely used in image recognition, speech processing, natural language processing, and other fields. Compared to traditional linear models, MLPs leverage multilayer structures and nonlinear activation functions to capture complex patterns and relationships in data, offering superior representational capacity and predictive performance.

Activation Functions Activation functions are crucial components in an MLP, introducing nonlinearity into the network, which allows the MLP to handle complex nonlinear problems. Common activation functions include Sigmoid, Tanh, and rectified linear unit (ReLU).

1. **Sigmoid.** The Sigmoid function maps input values to a range between 0 and 1, suitable for handling probabilities. However, the Sigmoid function can suffer from the vanishing gradient problem for very large or very small input values, impacting the training of deep networks. Its formula is as follows:

$$\sigma(x) = \frac{1}{1 + e^{-x}} \tag{2.9}$$

2. **Tanh.** The Tanh function, an extension of the Sigmoid, maps input values to a range between -1 and 1. The Tanh function has a zero-centered output, leading to faster convergence. However, it also suffers from the vanishing gradient problem:

$$\tanh(x) = \frac{e^x - e^{-x}}{e^x + e^{-x}} \tag{2.10}$$

3. **ReLU.** The ReLU [1] (rectified linear unit) function is currently the most widely used activation function due to its simplicity and effective training performance. ReLU outputs the input value if it is positive; otherwise, it outputs zero, effectively mitigating the vanishing gradient problem. However, ReLU can cause "dead" neurons, where neurons stop updating during training. Its formula is as follows:

$$\text{ReLU}(x) = \max(0, x) \tag{2.11}$$

Other variants like Leaky ReLU, ELU, and Swish provide different advantages in specific scenarios. The choice of activation function can significantly impact the performance of the MLP (Fig. 2.1).

Structure The structure of an MLP includes an input layer, one or more hidden layers, and an output layer. The number of neurons in each layer can be adjusted based on the specific problem requirements. The input layer is the first component of the MLP, responsible for receiving external input data and passing it to the hidden layers. Each neuron in the input layer corresponds to an input feature, such as pixel values in an image, word vectors in text, or spectral features in speech. The number of neurons in the input layer is determined by the dimensionality of the input data.

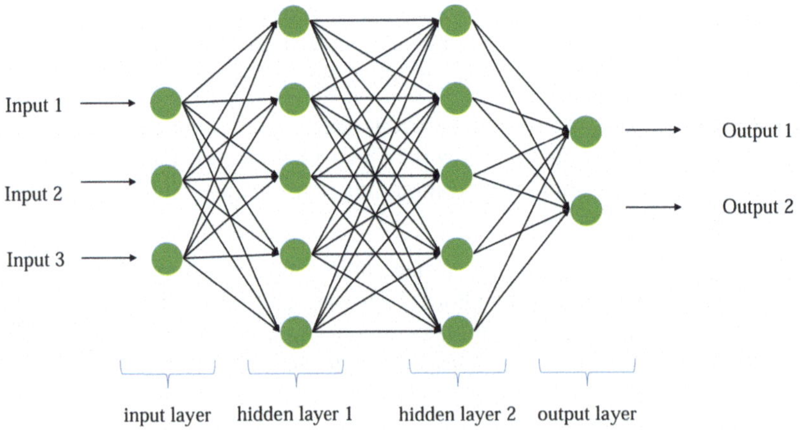

Fig. 2.1 Schematic diagram of multilayer perceptron network (MLP) architecture

For instance, a 28 × 28 grayscale image would have 784 neurons, each representing a pixel value. Neurons in the input layer do not process the input data; they simply transmit it to the next layer, the first hidden layer. The output of the input layer is essentially the input data itself, although preprocessing steps like normalization or standardization can adjust the data range and distribution to better suit subsequent layers. For example, in image processing tasks, pixel values are often normalized to the [0, 1] range to reduce computational load and accelerate convergence.

The hidden layers form the core of the MLP, consisting of one or more layers, each with multiple neurons. The primary function of the hidden layers is to perform feature extraction and nonlinear transformation on the input data, learning the complex patterns within the data. Each neuron in the hidden layer receives input from the previous layer, computes a weighted sum, adds a bias, and applies an activation function to produce the output.

The computation in the hidden layers involves three steps: weighted summation, bias addition, and activation function application. For the j-th neuron in the i-th hidden layer, the input is a weighted sum of the outputs from the previous layer:

$$z_j^{(i)} = \sum_k w_{jk}^{(i)} a_k^{(i-1)} + b_j^{(i)} \tag{2.12}$$

where $w_{jk}^{(i)}$ is the weight connecting the k-th neuron of the previous layer to the j-th neuron of the current layer, $a_k^{(i-1)}$ is the output of the k-th neuron in the previous layer, and $b_j^{(i)}$ is the bias for the j-th neuron in the current layer. The result $z_j^{(i)}$ is the net input.

Next, the net input $z_j^{(i)}$ is passed through an activation function to produce the output of the j-th neuron in the i-th layer:

$$a_j^{(i)} = \phi\left(z_j^{(i)}\right) \tag{2.13}$$

The activation function ϕ can be Sigmoid, Tanh, ReLU, etc. The nonlinearity introduced by the activation function allows the hidden layers to learn and represent the complex patterns and relationships in the data, enhancing the MLP's modeling capacity.

The number of hidden layers and the number of neurons in each layer are critical hyperparameters of the MLP. Increasing the number of hidden layers and neurons can enhance the model's representational power but also increases computational complexity and training time. Optimal hidden layer configuration is usually determined through cross-validation and experimentation.

The output layer is the final component of the MLP, generating the ultimate prediction. The number of neurons and the activation function used in the output layer depend on the specific task. For classification tasks, the number of neurons in the output layer corresponds to the number of classes, typically using the Softmax activation function to convert the outputs into a probability distribution.

The Softmax function is defined as follows:

$$\text{Softmax}(z_j) = \frac{e^{z_j}}{\sum_k e^{z_k}} \tag{2.14}$$

where z_j is the net input for the j-th neuron in the output layer. The Softmax function normalizes the outputs so that they sum to one, representing class probabilities.

For binary classification, the output layer may have a single neuron with a Sigmoid activation function to output a probability value:

$$\sigma(z) = \frac{1}{1 + e^{-z}} \tag{2.15}$$

For regression tasks, the number of neurons in the output layer matches the number of target variables, typically using a linear activation function to directly output the predicted values.

The design of the output layer significantly influences the MLP's predictive performance and interpretability. For multi-class classification tasks, using the Softmax activation function in the output layer allows for a straightforward interpretation of class probabilities, while for regression tasks, a linear output layer can provide precise predictions of continuous variables. Selecting the appropriate output layer structure and activation function is crucial for building a high-performance MLP.

Training Process The training process of an MLP employs the backpropagation algorithm, optimizing network weights by minimizing a loss function. The backpropagation algorithm is a gradient-based optimization method, consisting of forward propagation, loss computation, and backward propagation steps.

During forward propagation, input data is passed through each layer, applying weights and activation functions to produce predictions. The loss function then computes the error between the predicted and actual values. Common loss functions include mean squared error (MSE) for regression tasks, calculating the squared differences between predicted and actual values, and cross-entropy loss for classification tasks, measuring the difference between predicted probability distributions and actual distributions.

In the backward propagation phase, gradients of the loss function concerning each layer's weights are calculated using the chain rule. Starting from the output layer, the gradients for each weight are computed layer by layer backward. Finally, the weights and biases are updated using gradient descent to reduce the loss function. The learning rate, a crucial hyperparameter, determines the step size during weight updates. An appropriate learning rate can accelerate training and avoid local minima.

The MLP training process requires multiple iterations, repeating these steps until the loss function converges or a predefined number of epochs is reached. In practice, mini-batch gradient descent or batch gradient descent is often used to improve training efficiency. By dividing the training dataset into mini-batches, each update uses a subset of data, reducing computation and speeding up convergence.

Summary The MLP's primary advantage lies in its powerful nonlinear modeling capabilities and flexibility, making it suitable for various tasks due to its ability to handle complex nonlinear problems and relatively simple structure, which is easy to implement and train. In recent years, numerous studies have demonstrated the significant value of MLP in federated recommendation systems. Kalloori et al. [22] choose neural collaborative filtering (NCF) [16], a generalized MF based on MLP, to model user-item interactions among organizations. Further studies focus on communication efficiency [45], privacy security [20, 51], and customizing local models, such as in PFedRec [82], which fine-tunes global item embeddings and learns user-specific MLPs. Ning et al. [47] explore techniques to address performance degradation when sampling negatives from non-IID data, while Wu et al. [73] devise a hierarchical model to balance privacy and utility by embedding privacy-sensitive features locally and fusing public knowledge into private embeddings.

2.2.2 *Convolutional Neural Network*

Convolutional neural networks (CNNs) are a class of deep learning models extensively used in computer vision tasks, including image classification, object detection, and image segmentation. CNNs draw inspiration from the hierarchical processing mechanism of the biological visual system to extract multilayered features from input data. Their main advantage lies in their ability to automatically learn and extract significant features from data, reducing reliance on manual feature engineering. The convolutional neural network (CNN) composed of convolutional layers, pooling layers, and fully connected layers is prevalent in processing images, videos, and natural languages [31].

The architecture of a CNN typically consists of an input layer and multiple convolutional and pooling layers, followed by fully connected layers and an output layer. Each type of layer has a specific function within the network, collectively enhancing the model's performance and accuracy (Fig. 2.2).

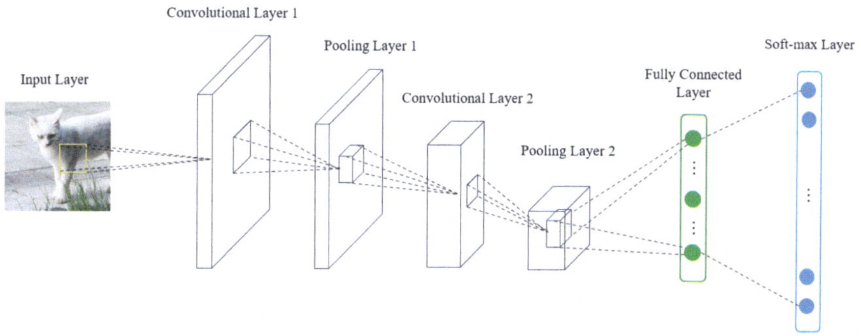

Fig. 2.2 Schematic diagram of convolutional neural network architecture

Input Layer The input layer is the first component of a CNN, responsible for receiving raw input data, typically represented as a three-dimensional matrix corresponding to the height, width, and number of color channels (e.g., RGB). For instance, a 32×32 RGB image has dimensions $32 \times 32 \times 3$. The input layer normalizes pixel values to the range [0, 1] or [-1, 1], aiding in stabilizing the training process and accelerating convergence. Additionally, the input layer may include preprocessing steps such as image scaling, centering, and normalization to ensure consistent scale and distribution of the input data throughout the training process. The output of the input layer is directly fed into the first convolutional layer. The primary function of the input layer in a CNN is to prepare data for efficient feature extraction in subsequent convolutional operations. Normalized data facilitates faster convergence and improved training stability, enabling CNNs to perform effectively on complex datasets.

Convolutional Layers Convolutional layers are the core building blocks of CNNs, responsible for extracting local features from input data. Each convolutional layer applies a set of learnable filters (or kernels) to the input data, performing convolution operations to capture local features such as edges, textures, and patterns. The convolution operation involves sliding the filter over the input data and computing the dot product between the filter weights and the input values within the receptive field, producing a feature map. The parameters of the filters are learned during training through backpropagation. The convolution operation can be mathematically described as follows:

$$Conv(I, K)(x, y) = \sum_{m} \sum_{n} I(x + m, y + n) K(m, n) \tag{2.16}$$

where I is the input image, K is the filter (kernel), and x and y are the spatial coordinates of the output feature map. This operation captures local dependencies and spatial hierarchies in the input data. Convolutional layers enable hierarchical feature extraction, progressively capturing features from low-level edges and textures to high-level object parts, enhancing the expressiveness and robustness of CNNs in processing complex data.

Following the convolution operation, an activation function is typically applied to introduce nonlinearity. The most commonly used activation function is the rectified linear unit (ReLU), defined as follows:

$$\text{ReLU}(x) = \max(0, x) \tag{2.17}$$

The ReLU activation function increases the model's nonlinear representation capacity by zeroing out all negative values while retaining positive values, helping to mitigate the vanishing gradient problem and accelerate the convergence of the training process. Due to its simplicity and efficiency, ReLU has become the most widely used activation function in CNNs. Other activation functions such as Leaky ReLU, ELU, and Swish may also be used in different scenarios to achieve better performance. The choice of activation function directly affects the training effectiveness and convergence speed of the model, making it a critical aspect of optimizing CNN performance based on specific tasks and data characteristics.

Pooling Layers Pooling layers, also known as subsampling layers, are responsible for reducing the spatial dimensions of the feature maps while retaining the most important information. This helps to decrease computational complexity and prevent overfitting. The most common types of pooling are max pooling and average pooling. Max pooling selects the maximum value within each pooling window, while average pooling computes the average value within each pooling window. The pooling operation can be mathematically described as follows for max pooling:

$$P(x, y) = \max_{i,j \in \text{pooling window}} F(x+i, y+j) \tag{2.18}$$

where F is the input feature map and P is the pooled feature map. Pooling significantly reduces the size of the feature maps while preserving critical features, reducing the number of parameters and computations, and enhancing the model's robustness to input variations such as translation and scaling. Pooling layers are typically applied after convolutional layers, allowing CNNs to progressively extract and compress feature representations of the input data. Pooling layers also exhibit some degree of noise resistance by selecting the maximum or average value within a local region, reducing the impact of noise in the input data. This enables CNNs to maintain high performance and stability when processing noisy data.

Fully Connected Layers Following a series of convolutional and pooling layers, high-level reasoning in the network is performed via fully connected layers. These layers resemble those in traditional neural networks, where each neuron is connected to every neuron in the previous layer. Fully connected layers combine the features extracted by convolutional layers to make predictions. The output of the final fully connected layer is typically passed through a Softmax activation function to produce class probabilities for classification tasks. The computation in the fully connected layers can be described as follows:

$$z_j = \sum_i w_{ji} a_i + b_j \tag{2.19}$$

where w_{ji} is the connection weight, a_i is the activation from the previous layer, b_j is the bias, and z_j is the output of the fully connected layer. This process allows fully connected layers to transform feature representations into final predictions. Fully connected layers have a large number of parameters because each neuron is connected to every neuron in the previous layer. To prevent overfitting, dropout layers are often introduced between fully connected layers, randomly dropping a fraction of the neurons' outputs to improve the model's generalization ability.

Training a CNN involves optimizing the network's parameters to minimize a loss function that measures the difference between the predicted output and the true labels. The training process includes the following steps:

1. **Forward Propagation:** The input data is passed through the network, layer by layer, to compute the output predictions.
2. **Loss Calculation:** The loss function (e.g., cross-entropy loss for classification) calculates the error between the predicted and true labels.
3. **Backward Propagation:** The gradients of the loss function with respect to each layer's weights are computed using the chain rule. This process starts from the output layer and proceeds backward through the network.
4. **Parameter Update:** The network's weights and biases are updated using an optimization algorithm such as stochastic gradient descent (SGD) or Adam to minimize the loss function.

The training process involves multiple iterations, each consisting of several epochs over the training dataset. The goal is to learn a set of parameters that generalize well to unseen data. In practice, mini-batch gradient descent or batch gradient descent is often used to improve training efficiency and stability. These steps enable CNNs to progressively learn useful features from the input data and make accurate predictions and classifications. To further enhance the model's performance and generalization ability, various regularization techniques such as dropout, weight decay, and data augmentation may be employed during training. These techniques effectively prevent overfitting and improve the model's performance on the test set.

Convolutional neural networks (CNNs) exhibit outstanding performance in handling complex data, with numerous notable advantages. Firstly, the local connectivity and weight-sharing mechanisms of convolutional layers enable CNNs to efficiently capture local features in input data while reducing the number of parameters. Secondly, pooling layers provide translation invariance, allowing the network to recognize objects regardless of their position. Additionally, CNNs can automatically learn hierarchical features, progressively building an understanding of input data from low-level to high-level features, excelling in complex tasks.

In federated recommendation systems, CNNs have demonstrated significant potential and broad applicability. They can model complex user-item interactions, enhancing the accuracy and personalization of recommendations. By training CNN models in a distributed manner across multiple devices, user privacy can be effectively protected while improving the performance of the recommendation system. The application of CNNs in federated learning primarily focuses on optimizing communication efficiency and privacy protection, jointly training more precise and robust recommendation models by aggregating data from multiple participants. Chen et al. [12] combine CNN with word embedding to model user and video representations from local textual information (e.g., user profiles and video description in document). Then, convoluted features are fed into PMF for predicting users' ratings on videos. Inspired by CASER [64], Zhu et al. [50] attempt to mine sequential patterns of user behaviors. Different from images and texts, convolutional kernels with horizontal and vertical shapes are applied to the embedding matrix of short-term interacted items.

2.2.3 Graph Neural Network

The graph neural network (GNN) is a powerful backbone for modeling graph structure data in manifold domains [36, 74]. GNNs are capable of capturing the intricate dependencies between nodes and edges, making them highly effective in applications such as social network analysis, recommendation systems, and molecular chemistry. At its core, a GNN leverages the structure of a graph to perform convolutions over nodes and their neighborhoods, effectively learning representations that capture both local and global graph information. This ability to model relational data and propagate information across the graph makes GNNs particularly suitable for tasks that involve complex and interconnected data structures.

When graph neural networks (GNNs) are applied to federated learning (FL), particularly in cross-user federations, each client only maintains an ego-network consisting of the current user node and the interacted item nodes. This localized structure inherently restricts the propagation of graph signals over high-order paths, thereby limiting the overall performance of the model. Addressing the challenge of capturing high-order connectivity among subgraphs distributed across clients necessitates the development of effective local graph expansion (LGE) strategies. An efficient LGE strategy would incorporate additional relevant nodes and edges, thereby enhancing the local graph structure and facilitating improved information flow.

Considering both LGE and graph heterogeneity, Table 2.3 provides a comparison of existing GNN-based HFedRec studies. These studies highlight various LGE approaches and address the challenges posed by graph heterogeneity, demonstrating the potential and complexity of implementing GNNs within a federated learning framework. FGC [80] and FL-GMT [28] are defined in the cross-organization federation, which naturally mitigates the constraints imposed by over-sparse local graphs. By aggregating graph information across organizations, these approaches enhance the model's ability to capture more comprehensive relationships between nodes, thus improving the recommendation performance.

Some studies [28, 38, 42] assume that social links are locally available. The works by Li et al. [28] and Liu et al. [38] design unique attention mechanisms to weigh the importance of user-item and user-user pairs. This selective attention mechanism allows the model to prioritize more relevant interactions, leading to more accurate recommendations. To alleviate local computation pressure, Ma et al. [42] assign a user model to each client and an item model to the server, respectively. This division of labor not only reduces the computational burden on individual clients but also ensures that item-related computations, which can be more data-intensive, are handled by the more capable server infrastructure.

Table 2.3 Comparison of GNN-based HFedRec from graph heterogeneity and local graph expansion

Publications	U/O	HN	HL	LGE	Model
FedPerGNN [72]	U	–	Social	Search	Frame
FedGR [42]	U	–	Social	–	Att
PerFedRec [40]	U	–	Social	Search	Frame
GPFedRec [81]	U	–	Social	Global	Frame
PerFedRec++ [41]	U	–	Social	Search	Frame
FedHGNN [75]	U	Multi	Multi	Global	Att
FGC [80]	O	–	–	–	Conv
CdFed [86]	U	–	–	Cluster	Frame
FL-GMT [28]	O	–	Social	–	Att
SemiDFEGL [54]	U	–	–	Cluster,P2P	Frame
FeSoG [38]	U	–	Social	–	Att
DGREC [89]	U	–	Social,Hyper	Global,P2P	Conv,Att

U/O: U, cross-user federation; O, cross-organization federation.
HN (heterogeneous nodes except users and items): Multi, multi-type nodes.
HL (heterogeneous links): Multi, multi-type links; Social, social links between users.
Hyper: Heterogeneous hyperedge connecting a subset of nodes [3];
LGE (local graph expansion): Search, search for neighbors based on matching mutual information.
Cluster, cluster of similar clients; Global, global graph reconstruction; P2P, peer-to-peer collaboration between clients.
Model: Frame, GNN-agnostic framework; Att, graph attention network (GAT).
Conv, graph convolutional network (GCN)

In privacy-sensitive scenarios of industrial IoT, Yin et al. [80] recommend suitable IoT services for local devices by a federated graph convolutional network. This approach leverages the decentralized nature of federated learning to ensure that sensitive data about IoT device usage remains local, thus preserving privacy while still benefiting from global model updates.

Without the assumption of prior social knowledge, Wu et al. [72] introduce a third party to search for similar users with co-interacted items, enriching the local graphs with additional relevant nodes. Luo et al. [40, 41] follow this practice to enhance local graphs with neighbors. PerFedRec++ goes further by considering client sampling and noise injection from privacy-preserving techniques as data augmentation, adopting a contrastive self-supervised task to pretrain models. This approach not only enriches the data available for training but also improves the robustness and generalization of the resulting models.

Both [40, 41] employ a clustering method to improve communication efficiency, detailed in Sect. 4.1. Another innovative idea is to privately reconstruct a pseudo-global graph connecting data islands. Wang et al. (GPFedRec [81]) reconstruct a social network in the server through user-user relevance computed by uploaded item embeddings. The server-side GNN then aggregates social-guided item embeddings to compensate for user-user collaborative signals, thereby enhancing the recommendation accuracy. Wang et al. (FedHGNN [75]) perform meta-path-based recommendation and generalize graph heterogeneity to multi-type nodes and links. Given public knowledge (e.g., item types) and perturbed local interactions, the server reconstructs a global heterogeneous information network (HIN) only once and then sends corresponding meta-paths back to clients. This method leverages the rich context provided by heterogeneous information to improve the recommendation model's understanding of complex user-item interactions.

In a decentralized federated setting, Zhao et al. (DGREC [89]) build an item hypergraph with tag information and perform an interest pooling operation to model user preference. Clients reconstruct social networks referring to [25] and share update gradients guided by social links. This approach enhances the robustness of the recommendation model by incorporating diverse sources of information and enabling more comprehensive user preference modeling. Moreover, Chen et al. [54] decide which neighbors a client can communicate with through a clustering method. According to uploaded local ego-network representations, the server clusters clients into groups, and P2P parameter passing is also constrained at the graph level. Similarly, after the server clusters clients by uploaded model weights, Liu et al. [86] allow clients to pass update gradients to intra-cluster neighbors. This hierarchical approach to communication and model updating enhances the efficiency and scalability of federated learning systems by reducing the communication overhead and ensuring more targeted information exchange.

In conclusion, the integration of GNNs with federated learning frameworks presents a promising avenue for enhancing recommendation systems. By effectively capturing high-order connectivity and addressing graph heterogeneity, GNN-based HFedRec approaches can significantly improve the quality and accuracy of recommendations while preserving user privacy. The ongoing research in this area continues to explore innovative strategies to overcome the inherent challenges, paving the way for more robust and scalable federated recommendation systems.

2.2.4 Autoencoders Network

The autoencoder network (AE) is an unsupervised learning model. An encoder maps the input data into a feature vector z, and a decoder attempts to reconstruct input data according to z. Then, z is regarded as a salient feature for subsequent recommendation tasks [83]. The fundamental architecture of an autoencoder network includes three main layers: the input layer, hidden layers, and the output layer. The input layer receives raw data, which can be images, text, or numerical data, with dimensionality matching the input data. The hidden layers facilitate the encoding and decoding processes. The encoder progressively reduces the dimensionality of the input data to extract essential features, creating a compressed representation. Conversely, the decoder expands this compressed representation back to the original dimensionality of the input data. The bottleneck layer, located between the encoder and the decoder, contains the feature vector z, representing the most compact and

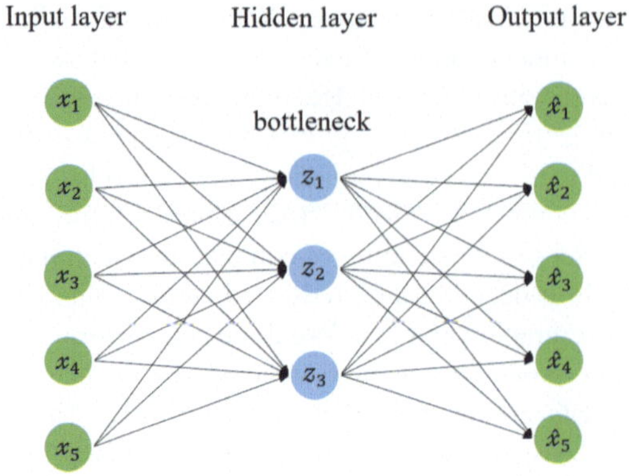

Fig. 2.3 Schematic diagram of autoencoder network architecture

informative depiction of the input data. The output layer's purpose is to reconstruct the input data from z, ensuring the reconstructed data retains the original format (Fig. 2.3).

Several autoencoder variants have been developed to address specific challenges and enhance performance, including variational autoencoders (VAEs) [24], denoising autoencoders (DAEs) [67], sparse autoencoders, and contractive autoencoders (CAEs) [56]. VAEs introduce probabilistic elements into the encoding process, mapping input data to a distribution over the latent space rather than a single feature vector. This approach enables the generation of new data samples by sampling from the learned distribution, making VAEs particularly valuable in generative tasks such as image and text generation. DAEs improve the robustness of learned features by training the model to reconstruct the original data from corrupted input versions, enhancing the model's ability to generalize and capture meaningful patterns. Sparse autoencoders introduce a sparsity constraint on hidden layers, promoting the learning of efficient and sparse representations by activating only a limited number of neurons. CAEs incorporate a regularization term in the loss function, penalizing the sensitivity of hidden layer activations to minor changes in the input data, fostering the learning of robust features less sensitive to noise. Training an autoencoder involves optimizing both the encoder and decoder parameters to minimize the reconstruction error, which quantifies the discrepancy between the input and reconstructed data. Common loss functions include mean squared error (MSE) and binary cross-entropy. The training process typically includes forwarding the input data through the encoder to derive z, passing z through the decoder to reconstruct the data, computing the reconstruction error using a chosen loss function, calculating gradients via backpropagation, and updating model parameters through optimization algorithms such as stochastic gradient descent (SGD) or Adam.

Autoencoders offer several advantages, including the ability to perform unsupervised learning, feature extraction, and dimensionality reduction. They do not require labeled data, making them suitable for tasks where data labeling is expensive or impractical. Autoencoders can autonomously learn valuable features from the data, applicable to various downstream tasks. Additionally, they can reduce data dimensionality, facilitating easier visualization and analysis.

By leveraging the advantages of autoencoders, numerous studies have applied them to recommendation systems, achieving significant results. Chen et al. [19] employ two variational autoencoders (VAEs) [59] in both server and client sides for sequential recommendation. To avoid high variance and posterior collapse, the client-side VAE discretizes temporally encoded item sequences through a vector quantization technique. The server-side VAE takes uploaded encoder weights as input and models their semantic similarity. Within-cluster encoders are aggregated after k-means clustering. Wang et al. [49] define concept-stock recommendation problem on public stock reports, private forum comments, and experts' knowledge. Transformer graph layers [61] are acted as an encoder to learn fused embeddings of concepts and stocks, while linear layers aim to reconstruct the two source embeddings.

2.2.5 Mixture Model

Mixture models (MMs) represent an advanced class of models that integrate multiple types of neural network building blocks to harness the strengths of various architectures. These models are designed to handle complex data structures and relationships by combining diverse neural components, such as convolutional layers, recurrent layers, and attention mechanisms. The core principle of mixture models is to employ the most suitable neural architectures for different parts of the data, facilitating robust and efficient learning.

In a mixture model, various neural components address specific tasks within the overall model architecture. For instance, convolutional layers excel at extracting local patterns from spatial data, while recurrent layers, such as LSTMs and gated recurrent units (GRUs), capture temporal dependencies in sequential data. Attention mechanisms enhance the model's capacity to focus on the most relevant parts of the input, thereby improving overall performance by emphasizing critical information. By integrating these diverse components, mixture models effectively handle heterogeneous data sources and complex dependencies.

The flexibility of mixture models allows them to be tailored to specific applications and data types, making them highly versatile. They can be adapted to various tasks, such as classification, regression, and clustering, by selecting and combining appropriate neural network components. This adaptability is particularly beneficial in recommendation systems, where the nature of the data and the requirements of the task can vary significantly.

Mixture models offer several advantages, including versatility, enhanced performance, robustness, and scalability. They can handle different types of data and tasks by combining suitable neural network components, achieving superior performance compared to single-type models. Their integration of various neural components enables them to handle complex and noisy data more effectively. However, mixture models also present challenges, such as increased complexity, higher computational costs, and the need for careful hyperparameter tuning. The combination of multiple components can complicate the design and implementation of mixture models, making them more resource-intensive and harder to interpret compared to simpler models.

Mixture models have been effectively applied in recommendation systems to improve the quality of recommendations by leveraging the strengths of different neural network architectures. Luo et al. [53] aim at privacy-preserving news recommendations, comprising two core modules: (i) The user module is a self-attention-extended LSTUR model [15] and learns user embeddings from historical user behaviors. (ii) The news module learns news embeddings from news titles and embodies word embedding, attention, and convolutional layers. Xu et al. [79] further put the heavyweight news model on the server side and the lightweight user model on the client side to offload clients' computational overhead. To learn preference evolution with subtle temporal dependency, Zhu et al. [26] combine low-rank tensor projection and GRU layers to model long- and short-term interests, respectively. He et al. (CPF-POI [78]) first locally remove location noises within historical trajectories. Then embedded location and timestamp records are attentively fused and fed into GRU layers considering the spatial-temporal interval. Similarly affected by the sparsity of trajectories, Zhang et al. [87] attentively fuse historical recent and periodic preference with the awareness of time interval. Relying on the interpretable inference and structural semantics of knowledge graphs, Wei et al. [71] inject item knowledge distilled by GCN into transformer-based sequential recommendation.

In conclusion, mixture models provide a versatile and powerful framework for recommendation systems by integrating various neural network components tailored to specific data characteristics and requirements. Their ability to handle diverse data types and dependencies makes them particularly effective in delivering personalized and context-aware recommendations.

2.2.6 Summary and Challenges

Despite being validated on specific models, some works [40, 50, 80] tend to be generic frameworks suitable for a broader range of neural networks. Meanwhile, Peng et al. [49], Liu et al. [38], and Ye et al. [78] devised for the horizontal federation can also work well in the centralized setting. However, computation overhead for high-dimensional model weights and embedding matrices are more expensive than statistical machine learning approaches.

One specific challenge is **how to design federated models that concentrate intensive computing on the server side**. Yi et al. [79] and Ma et al. [42] assign the burdensome task of item representation to the server, which shares a similar idea to split learning [65]. This approach helps reduce the computational burden on client devices, making the federated learning process more efficient and feasible for devices with limited resources. Knowledge distillation [29] also enables clients to own tinier models, introduced in Sect. 4.1. This technique involves distilling knowledge from a larger, more complex model into a smaller, more manageable one for the clients, thereby easing the computational demands on individual clients.

For sequential and graphical data, another challenge is **how to enhance feature diffusion across local sequences and graphs**. The vanilla federated averaging strategy [43] is insufficient to explicitly capture common sequential and graphical patterns among clients. This limitation necessitates the development of advanced federated models that can effectively handle the intricacies of sequential and graphical data. More federated models are expected to highlight the computation pressure and data characteristics, rather than copy the federated procedure directly.

Furthermore, ensuring efficient feature propagation and aggregation in GNNs within federated learning frameworks is critical. Techniques such as local graph expansion (LGE) and attention mechanisms need to be refined and integrated to improve the performance of GNN-based federated models. Privacy-preserving mechanisms must also be robust enough to handle the complexity of these advanced models without compromising on data security.

In summary, while deep learning approaches in federated learning show great promise, they face several challenges, including computational overhead, efficient feature diffusion, and data heterogeneity. Continued research and innovation are required to develop federated models that are not only computationally efficient but also capable of leveraging the unique characteristics of sequential and graphical data. Addressing these challenges will be crucial for advancing federated learning and its applications across various domains.

2.3 Meta-Learning

Machine learning- and deep learning-based HFedRec have combined federated learning with many kinds of recommendation algorithms, but there are still two issues:

(1) Different users tend to present personalized behavior patterns especially in cross-user HFedRec. For instance, some have a long-term preference for digital products, while others prefer daily necessities according to their short-term needs.
(2) All clients' recommendation tasks are treated as single learning problem to train a global model performing well for most of them. The global model may ignore uncommon preference.

Therefore, meta-learning comes into the field of HFedRec. The nucleus concept of meta-learning is "learning to learn," and the emphatic problem is "how to learn a model faster and better in the near future." From the view of task distribution, each client's user-item interaction pattern can be regarded as a training *task*. Meta-learning intends to obtain a general-purpose learning algorithm that quickly adapts to a new scenario within a few local training steps [17]. In the following subsections, we will describe two typical meta algorithms of model-agnostic meta-learning (MAML) [14] and Reptile [46] explored in HFedRec.

2.3.1 MAML-Based Approaches

Meta-learning, particularly model-agnostic meta-learning (MAML), has been gaining significant traction in the field of federated learning due to its capability to quickly adapt models to new tasks with minimal data. MAML is designed to optimize the model parameters such that a small number of gradient updates can lead to rapid learning on a new task. This is particularly useful in federated learning scenarios where data is distributed across multiple clients, each with its unique distribution.

In the case of MAML, local preference data are required to be divided into support set and query set. The support set first updates model parameters θ_0, and then the query set points to the real gradient descent direction:

$$\begin{aligned} \theta_i &= \theta_0 - \eta_1 \nabla_\theta L_{D_i^s}(f_{\theta_0}) \\ \theta_0 &= \theta_0 - \eta_2 \frac{1}{|T|} \sum_{T_i} \nabla_\theta L_{D_i^q}(f_{\theta_i}) \end{aligned} \tag{2.20}$$

where D_i^s and D_i^q are the support set and query set of i-th task T_i, η_1 and η_2 are learning rates at different training stages, and θ_i denotes the updated model parameters by support set D_i^s. When MAML meets federated learning, clients serve as *tasks*, and the number of tasks $|T|$ also refers to the number of sampled clients. The summation of gradients $\nabla_\theta L_{D_i^q}(f_{\theta_i})$ is computed in the server. FedMeta [7] first defines the primary procedure:

- **Step 1.** The server initializes model parameters θ_0 and samples a batch of clients C (equivalent to tasks $\{T_i, i \in [C]\}$) to start the federated training process.
- **Step 2.** Selected clients C download model initialization parameters θ_0 from the server and then utilize the support set D_i^s to train local models by one (or more) gradient descent step.
- **Step 3.** After completing local training steps in parallel, clients send test gradients computed by the query set D_i^q to the server.
- **Step 4.** The server aggregates test gradients uploaded by clients and updates model initialization parameters θ_0. Then the server distributes initialization parameters to clients for the next communication round.
- **Step 5.** Repeat steps 2–4 until convergence.

Besides, Li et al. [88] combine first-order MAML with bidirectional gated recurrent units (GRUs) and attention layers. It considers two different user groups participating in HFedRec: public users who agree to share personal data and private users who disagree to share. MetaMF [34] is a meta matrix factorization that consists of a collaborative memory (CM) module, a meta recommender (MR) module, and a rating prediction (RP) module. The RP module is deployed in local devices for predicting ratings of items. CM and MR modules, deployed in the server, aim to encode users' behaviors and generate parameters of private RP modules, respectively.

Meta-learning, and particularly MAML, addresses two primary issues in HFedRec: (1) the need to accommodate personalized behavior patterns among users and (2) the challenge of creating a global model that performs well across diverse user preferences. In cross user HFedRec, different users exhibit unique behavior patterns. For example, some users may have a long-term preference for digital products, while others may prefer daily necessities based on short-term needs. Treating all clients' recommendation tasks as a single learning problem to train a global model can lead to the neglect of uncommon preferences.

The central concept of meta-learning is "learning to learn," which focuses on how to train a model faster and better in future scenarios. From the perspective of task distribution, each client's user-item interaction pattern can be regarded as a training task. Meta-learning aims to develop a general-purpose learning algorithm that can quickly adapt to new scenarios with a few local training steps. This capability is particularly beneficial in federated learning, where client data distributions can vary significantly.

In the context of HFedRec, MAML-based approaches such as FedMeta, Fed4Rec, and MetaMF provide robust frameworks for addressing the heterogeneity of client data and enabling rapid adaptation to new user preferences. These methods highlight the potential of integrating meta-learning with federated learning to enhance the personalization and effectiveness of recommendation systems.

2.3.2 *Reptile-Based Approaches*

Compared with MAML, Reptile [46] consumes less computational resources without twice derivative calculations and the separation of support set and query set. The key distinction consists in the update of model initialization parameters:

$$\theta_0 = \theta_0 + \eta \frac{1}{|T|} \sum_{T_i} (\theta_i - \theta_0) \tag{2.21}$$

Reptile directs model initialization parameters to the final updated parameters learned from tasks. Reptile, as an optimization-based meta-learning algorithm, offers an efficient alternative to MAML by simplifying the meta-training process. Unlike MAML, which requires computing second-order derivatives, Reptile involves first-order optimization, making it computationally less intensive. This reduction in computational complexity is particularly advantageous in federated learning settings, where resource constraints are prevalent. The core concept of Reptile is to iteratively adjust the initialization parameters toward the parameters of models trained on various tasks. By doing so, Reptile aims to find an optimal initialization that can rapidly adapt to new tasks with minimal updates. This approach eliminates the need for a distinct separation between support and query sets, as required by MAML, further streamlining the training procedure.

In the context of federated learning, Reptile's efficiency and simplicity make it particularly well-suited. Federated learning environments often involve numerous clients with varying amounts of data and computational resources. Reptile's capability to perform meta-learning without extensive computational power makes it an ideal candidate for such distributed systems.

Wang et al. in their work (DP-PrivRec [70]) implement the meta update by Reptile to quickly adapt inactive users to local preference patterns. DP-PrivRec leverages Reptile's meta-update mechanism to quickly adapt to the preference patterns of inactive users. This method enhances recommendation accuracy for users who may not frequently interact with the system, ensuring that the model remains personalized and effective even for less active users. Such rapid adaptation is crucial for maintaining user engagement and satisfaction in recommendation systems.

Using Taylor series expansion to approximate the Reptile-style aggregation in the server, Wang et al. [9] additionally give a theoretical proof that the meta update has the ability of global model generalization. FedISFedIS extends Reptile's application in federated learning by utilizing Taylor series expansion to approximate the aggregation of model updates at the server. This method not only preserves Reptile's efficiency but also provides a theoretical foundation for the global generalization capability of the meta-updated model. Effective generalization across diverse clients is essential for developing robust federated learning models that perform well in heterogeneous and dynamic environments.

In summary, Reptile-based approaches offer a practical and efficient framework for meta-learning in federated settings. By reducing computational requirements and simplifying the training process, Reptile facilitates the rapid adaptation of models to new tasks and user preferences. The successful implementations in DP-PrivRec and FedIS demonstrate Reptile's potential in enhancing the personalization and generalization of federated learning systems. These advancements underscore the significance of Reptile as a valuable tool for developing scalable and effective recommendation systems in federated learning environments.

2.3.3 Summary and Challenges

Meta-learning does not require a global model to reach optimal performance of all clients at once, and local models can converge within fewer training epochs, of which both make it possible to build efficient HFedRec. Therefore, a latent challenge is **how to further combine efficiency-enhancing approaches with the meta paradigm**. For instance, Wu et al. [84] and Xu et al. [79], mentioned in

Sects. 2.1 and 2.2, have made efficiency improvement in their own way. It remains a surprise when meta-learning combines with existing efforts.

Furthermore, since federated meta-learning aims at a fast adaptation from the global model to local heterogeneous tasks, it is natural to promote another subject named *personalized federated learning* (PFL) [63]. One advantage of PFL is to handle non-IID data varying in size and distribution. Thus, another significant challenge is **how to incorporate personalization strategies in PFL**. Personalized federated learning strategies can tailor the global model to better fit the unique preferences and behaviors of individual users, thereby enhancing the overall performance of the recommendation system.

Another critical challenge is ensuring the scalability of federated meta-learning systems. As the number of clients and the volume of data increase, the computational and communication overhead can become significant. Efficient algorithms and communication protocols need to be developed to handle large-scale federated learning scenarios. By addressing these challenges, federated meta-learning can become a powerful tool for building efficient, personalized, and scalable recommendation systems that respect user privacy and deliver high-quality experiences.

2.4 Reinforcement Learning

Reinforcement learning (RL) is dedicated to the ongoing interaction between the agent and the environment, rather than learning pre-defined tasks for various scenarios [48]. In each agent-environment interaction, the agent determines the next action based on the observed state of the environment. The multi-armed bandit (MAB) problem is a classic example of RL:

(1) The *agent* is a player (i.e., RecSys), and the *environment* is a bandit (i.e., user) with K arms (i.e., items). At each time slot, the player decides which arm to pull from the set of K arms.
(2) Each pulled arm has different probability distribution for gaining coins (i.e., rewards), while the player aims to know which arm can maximize the total reward during the playing process.

Consequently, RL-based RecSys can make accurate recommendations in dynamic environments, irrespective of whether the current scenario is represented in the datasets. We categorize existing RL-based HFedRec into multi-agent and single-agent distributions to detail their solutions.

Reinforcement learning involves several essential components beyond the agent and environment, including state, action, policy, and rewards, which are integral to the learning process. The *state* represents the current condition of the environment as perceived by the agent. In recommendation systems (RecSys), the state encompasses features describing the user's current interaction with the system, such as the user profile, past interactions, preferences, and contextual information like time of day or location. This state information enables the RecSys to understand the user's needs and preferences at any given moment, allowing for tailored recommendations.

An *action* in RL refers to the decisions or moves made by the agent in response to the current state. In RecSys, actions correspond to the recommendations or items suggested to the user. The agent (RecSys) selects an item to recommend based on the current state (user's preferences and context) to maximize the overall reward (user satisfaction and engagement). The choice of action is critical as it directly influences the rewards obtained and future states of the environment.

The *policy* in RL defines the strategy used by the agent to determine its actions based on the observed states. Policies can be deterministic, where a specific action is chosen for a given state, or probabilistic, where actions are selected according to a probability distribution. In RecSys, the policy dictates how recommendations are generated. An effective policy ensures that the RecSys learns and adapts to the user's changing preferences, providing increasingly personalized and relevant recommendations.

Rewards are the feedback received by the agent following an action, indicating the success or failure of that action in achieving the desired outcome. In the context of RecSys, rewards may include user interactions such as clicks, likes, purchases, or other forms of engagement. The objective of the RecSys is to maximize cumulative rewards, thereby enhancing user satisfaction and retention through more effective recommendations.

In RL-based recommendation systems, we distinguish between multi-agent and single-agent distributions. In single-agent systems, a single recommendation agent interacts with the user environment, aiming to optimize recommendations based on individual user interactions. This approach is straightforward but may not fully capture the complexity of environments with multiple interacting users or agents. Multi-agent systems involve multiple agents interacting within the same environment, each potentially representing different recommendation strategies or focusing on various aspects of the user experience. These agents can collaborate or compete to optimize overall recommendation quality. Multi-agent systems are more complex and require sophisticated coordination mechanisms, but they can deliver more robust and diverse recommendations.

By comprehending these key components and their roles within RL, we can design and implement recommendation systems that adeptly adapt to users' dynamic needs and preferences, ultimately enhancing the overall user experience.

2.4.1 Multi-Agent Distribution

The horizontal federation places each client in an independent *environment*, and clients cannot interfere with each other directly. Thus, local models can be regarded as individual *agents* and play the same set of K arms. The server takes charge of aggregating observations uploaded by clients. This decentralized approach is crucial in ensuring that each agent operates based on its local environment while contributing to a global model without direct interference from other agents. In this setup, the server-client federated process aims to find the optimal arm with high probability, ensuring efficient decision-making across the network.

One of the key advantages of multi-agent systems in reinforcement learning is their ability to effectively coordinate and collaborate in a distributed environment. Each agent operates independently in its local environment and makes decisions based on local observations, thereby enhancing the overall robustness and adaptability of the system. This approach not only reduces the risk of single points of failure but also provides greater flexibility in responding to dynamic environmental changes. Fed2-UCB [60] first defines a systematic federated multi-armed bandit (MAB) framework experimented on recommendation tasks. The generic steps are the following:

- **Step 1.** All clients initialize active arms K_{act} (i.e., candidate items), and the server samples a batch of clients C to start federated training steps.
- **Step 2.** Selected clients pull each active arm for a constant number of times. Sample means $\{\overline{\mu}_{c,k} | c \in C, k \in K_{\text{act}}\}$ are updated based on users' feedback and transmitted to the server. *(Sub-Round I)*
- **Step 3.** The server updates the global σ-sub-Gaussian distribution with aggregated sample means $\sum_{c \in C} \overline{\mu}_{c,k} / |C|$. Then, double upper confidence bounds (UCBs) [4] are utilized to determine the confidence bound and suboptimal arms K_{eli} to be eliminated.
- **Step 4.** Selected clients download the elimination arm set from the server to decrease active arms $K_{\text{act}} = K_{\text{act}} \backslash K_{\text{eli}}$. *(Sub-Round II)*
- **Step 5.** Repeat steps 2 to 4 until one active arm is left.

The double UCBs aim to handle uncertainty brought by client sampling and arm sampling. The former refers to the possibility that the global bandit model may not fairly select the optimal arm due to only part clients incorporated in one communication round, while the latter addresses the common exploration-exploitation dilemma of reinforcement learning (RL) [85]. The exploration-exploitation tradeoff is a fundamental challenge in RL, where the agent must balance between exploring new actions to discover their effects and exploiting known actions that yield high rewards. The double UCBs help to systematically manage this trade-off by providing a robust mechanism for arm selection and elimination.

Moreover, Fed2-UCB has a simplified version named Fed1-UCB in a special case where the client sampling is excluded, and all clients join in the decision-making process. This simplification can be beneficial in scenarios where communication overhead and computational complexity need to be minimized, ensuring that the model remains efficient and scalable.

Similarly, He et al. [37] adapt federated MAB to aspect-based recommendations. In aspect-based recommendations, users can rate different aspects of an item, such as the price, performance, and battery life of a phone. This multifaceted rating system requires a nuanced approach to recommendation, where the model must account for various dimensions of user preferences. FPAE evaluates whether the model can effectively perform the arm-elimination phase with a single communication round when communication is expensive. This evaluation is critical in environments where communication resources are limited, and minimizing the number of communication rounds can lead to significant savings in time and bandwidth.

In reinforcement learning, particularly in multi-agent systems, the goal is to enable agents to learn and make decisions that maximize their cumulative rewards. In federated settings, where agents operate independently but contribute to a shared global model, it is essential to ensure that the learning process is efficient and effective. The approaches like Fed2-UCB and FPAE provide a structured methodology to achieve this by balancing exploration and exploitation and managing the communication constraints inherent in federated learning environments.

Furthermore, the ability of these models to adapt to dynamic environments and user preferences highlights their robustness and flexibility. As federated learning continues to evolve, the integration of advanced reinforcement learning techniques will be crucial in developing systems that can efficiently handle the complexities of real-world applications. The continued research and development in this area promise to enhance the capabilities of federated learning systems, making them more adaptive, scalable, and effective in various domains. The multi-agent system also provides scalability to the learning process. As the number of agents increases, the system can handle more tasks simultaneously, distributing the workload and ensuring that the system can scale to meet the demands of larger and more complex environments. This scalability is essential for the deployment of reinforcement learning systems in real-world applications, where the ability to manage and process vast amounts of data from numerous sources is a significant advantage.

The integration of multi-agent distribution in reinforcement learning, particularly through frameworks like Fed2-UCB and FPAE, provides a robust approach to decentralized decision-making in federated settings. By managing the uncertainties associated with client and arm sampling and addressing the exploration-exploitation trade-off, these frameworks enhance the efficiency and effectiveness of federated learning systems. These advancements underscore the importance of collaborative approaches in tackling complex problems in distributed environments, paving the way for more adaptive and scalable reinforcement learning applications.

2.4.2 Single-Agent Distribution

In this subcategory, the server serves as a global *agent*, and clients are regarded as the *environment*. Owing to the communication payload proportional to the size of items [2, 5], Khan et al. [23] propose a MAB-based optimization method named FCF-BTS to alleviate the payload pressure. FCF-BTS computes rewards based on submitted item gradients and leverages Bayesian Thompson sampling to formulate a sampling strategy. The central bandit model provides clients with an appropriate set of candidate items and minimally intervenes in local recommendation decisions.

Single-agent distribution frameworks in federated learning are characterized by a central server acting as the sole decision-making entity, or agent, which coordinates the learning process across multiple clients. This centralization simplifies the overall architecture by reducing the need for complex coordination mechanisms among multiple agents. The server aggregates data and updates from all clients, which are treated as part of the environment, to build and refine a global model. The primary advantage of this approach is its ability to streamline communication and computational processes. By centralizing control, single-agent frameworks can more effectively manage the aggregation of information, leading to more efficient and consistent updates to the global model. This centralized structure is particularly advantageous in scenarios where maintaining synchronization among distributed clients can be challenging.

The FCF-BTS method aims to optimize the selection of candidate items for recommendation while minimizing communication overhead. By using Bayesian Thompson sampling, the method dynamically adjusts the sampling strategy based on the observed rewards, leading to more efficient exploration and exploitation of the item space. This approach ensures that the central agent can provide high-quality recommendations with reduced communication costs, making it suitable for large-scale federated learning systems. This method underscores the importance of adaptive sampling strategies in reducing the strain on communication resources while maintaining the quality of recommendations.

Li et al. [87] adopt distillation-based reinforcement learning (RL) to enhance clustering capability in PFL-based HFedRec. The server maintains two clustering networks, one as a teacher providing guidance and the other as a policy requiring optimization. Rewards are measured by local model performance, and the policy network improves by maximizing rewards obtained from both the environment and the teacher network. This distillation-based approach allows the server to leverage the expertise of the teacher network to refine the policy network, leading to more effective clustering and better overall performance in personalized federated learning scenarios. The use of distillation in RL contexts underscores the potential for combining traditional machine learning techniques with modern RL strategies to achieve superior outcomes.

Single-agent distribution frameworks offer several advantages, including simplified communication protocols and centralized control, which can lead to more consistent and reliable model updates. However, they also face challenges such as potential bottlenecks at the central server and the need to balance the load across clients to prevent any single client from becoming a point of failure. Effective load balancing and efficient communication strategies are crucial for maintaining system performance and scalability. Furthermore, the centralized approach must be designed to handle varying levels of data quality and quantity across clients, ensuring that the global model remains robust and effective.

The single-agent approach allows for more straightforward implementation of advanced machine learning and deep learning techniques, as the central agent can directly incorporate complex algorithms and models without the need for extensive coordination among multiple agents. This capability is particularly beneficial in scenarios where sophisticated models are required to capture intricate patterns and relationships within the data. By centralizing the computational load, the server can employ more powerful computational resources and advanced optimization methods to enhance model performance.

Additionally, single-agent distribution frameworks can facilitate more streamlined deployment and maintenance of federated learning systems. Centralized updates and control make it easier to implement system-wide improvements and updates, ensuring that all clients benefit from the latest advancements in model development and optimization. This centralized management also simplifies the process of monitoring and debugging the system, leading to more efficient identification and resolution of issues.

Moreover, single-agent systems can enhance the security and privacy of the federated learning process. With a central entity managing the data aggregation and model updates, it is easier to enforce strict security protocols and monitor for potential breaches. This centralized oversight can ensure that sensitive data remains protected throughout the learning process.

In conclusion, the single-agent distribution in federated learning provides a streamlined and efficient framework for managing the learning process across distributed clients. By centralizing the decision-making and leveraging advanced optimization techniques like Bayesian Thompson sampling and distillation-based reinforcement learning, these frameworks can achieve high performance while minimizing communication overhead. As federated learning continues to evolve, single-agent distribution models will play a crucial role in enabling scalable and effective learning in diverse and dynamic environments. The ongoing advancements in this area promise to enhance the capabilities of federated learning systems, making them more adaptive, scalable, and capable of addressing complex real-world challenges.

2.4.3 Summary and Challenges

In personalized recommendation, the exploration-exploitation dilemma of reinforcement learning (RL) refers to the phenomenon of information cocoons or echo chambers [66], where individuals consistently receive content aligned with their existing interests and opinions. Multi-armed bandit (MAB)-based RL provides HFedRec with a fair strategy to address this polarization and balance exploration and exploitation. Multi-agent distribution frameworks mitigate this dilemma by allowing multiple agents to operate independently within their local environments while contributing to a shared global model, thereby enhancing the system's robustness and adaptability. Conversely, single-agent distribution frameworks centralize the decision-making process through a single global agent that coordinates the learning process across all clients, simplifying the architecture and improving overall efficiency.

Despite these advantages, certain items may be popular in one environment but fail to demonstrate their popularity in the broader federated context. Globally learned policies (e.g., upper confidence bound (UCB)) tend to be suboptimal in heterogeneous environments [21]. The primary challenge lies in **how agents can adaptively fit the global policy according to local environments**, akin to the propositions in personalized federated learning (PFL). Additionally, current practices predominantly focus on the MAB problem, presenting another challenge: **how to integrate the broader spectrum of RL**, including deep reinforcement learning [68] and reinforcement distillation [57], which are promising areas for exploration within HFedRec.

References

1. Agarap AF (2019) Deep learning using rectified linear units (relu). https://arxiv.org/abs/1803.08375
2. Ammad-Ud-Din M, Ivannikova E, Khan SA, Oyomno W, Fu Q, et al (2019) Federated collaborative filtering for privacy-preserving personalized recommendation system
3. Antelmi A, Cordasco G, Polato M, Scarano V, Spagnuolo C, Yang D (2023) A survey on hypergraph representation learning. ACM Comput Surv 56(1):1–38
4. Auer P, Cesa-Bianchi N, Fischer P (2002) Finite-time analysis of the multiarmed bandit problem. Mach Learn 47:235–256
5. Chai D, Wang L, Chen K, Yang Q (2021) Secure federated matrix factorization. IEEE Intell Syst 36(5):11–20
6. Chen C, Zhang J, Tung AK, Kankanhalli M, Chen G (2020) Robust federated recommendation system
7. Chen F, Luo M, Dong Z, Li Z, He X (2019) Federated meta-learning with fast convergence and efficient communication
8. Da'u A, Salim N (2020) Recommendation system based on deep learning methods: a systematic review and new directions. Artif Intell Rev 53(4):2709–2748
9. Ding X, Li G, Yuan L, Zhang L, Rong Q (2023) Efficient federated item similarity model for privacy-preserving recommendation. Inf Process Manag 60(5):103470
10. Dong Q, Liu B, Zhang X, Qin J, Wang B, Qian J (2022) Ranking-based federated poi recommendation with geographic effect. In International joint conference on neural networks
11. Du Y, Zhou D, Xie Y, Shi J, Gong M (2021) Federated matrix factorization for privacy-preserving recommender systems. Appl Soft Comput 111:107700
12. Duan S, Zhang D, Wang Y, Li L, Zhang Y (2020) Jointrec: a deep-learning-based joint cloud video recommendation framework for mobile IoT. IEEE Internet Things J 7(3):1655–1666
13. Duriakova E, Tragos EZ, Smyth B, Hurley N, Peña FJ, et al (2019) PDMFRec: a decentralised matrix factorisation with tunable user-centric privacy. In Proceedings of the 13th ACM conference on recommender systems, RecSys '19. ACM Press, New York, pp 457–461
14. Finn C, Abbeel P, Levine S (2017) Model-agnostic meta-learning for fast adaptation of deep networks. In Proceedings of the 34th international conference on machine learning. PMLR, New York
15. Ginker T, Lieberman O (2021) Lstur regression theory and the instability of the sample correlation coefficient between financial return indices. Econ J 24(1):58–82
16. He X, Liao L, Zhang H, Nie L, Hu X, Chua TS (2017) Neural collaborative filtering. In Proceedings of the Web Conference, WWW '17, pp 173–182. International World Wide Web Conferences Steering Committee, Republic and Canton of Geneva, CHE

17. Hospedales T, Antoniou A, Micaelli P, Storkey A (2022) Meta-learning in neural networks: a survey. IEEE Trans Pattern Anal Mach Intell 44(9):5149–5169
18. Hu Y, Koren Y, Volinsky C (2008) Collaborative filtering for implicit feedback datasets. In IEEE international conference on data mining, ICDM '08. IEEE Computer Society, USA, pp 263–272
19. Imran M, Yin H, Chen T, Nguyen QVH, Zhou A, Zheng K (2023) ReFRS: Resource-efficient federated recommender system for dynamic and diversified user preferences. ACM Trans Inf Syst 41(3):1–30
20. Jiang X, Liu B, Qin J, Zhang Y, Qian J (2022) FedNCF: federated neural collaborative filtering for privacy-preserving recommender system. In International Joint Conference on Neural Networks
21. Jin H, Peng Y, Yang W, Wang S, Zhang Z (2022) Federated reinforcement learning with environment heterogeneity. In Proceedings of The 25th international conference on artificial intelligence and statistics. PMLR, New York, pp 18–37
22. Kalloori S, Klingler S (2021) Horizontal cross-silo federated recommender systems. In Fifteenth ACM conference on recommender systems. ACM Press, New York, pp 680–684
23. Khan FK, Flanagan A, Tan KE, Alamgir Z, Ammad-ud din M (2021) A payload optimization method for federated recommender systems. In Fifteenth ACM conference on recommender systems. ACM Press, New York, pp 432–442
24. Kingma DP, Welling M (2022) Auto-encoding variational bayes. https://arxiv.org/abs/1312.6114
25. Krasanakis E, Papadopoulos S, Kompatsiaris I (2022) p2pgnn: A decentralized graph neural network for node classification in peer-to-peer networks. IEEE Access 10:34755–34765
26. Li L, Lin F, Xiahou J, Lin Y, Wu P, Liu Y (2022) Federated low-rank tensor projections for sequential recommendation. Knowl-Based Syst 255:109483
27. Li Q, Wei X, Lin H, Liu Y, Chen T, Ma X (2022) Inspecting the running process of horizontal federated learning via visual analytics. IEEE Trans Vis Comput Graph 28(12):4085–4100
28. Li Z, Bilal M, Xu X, Jiang J, Cui Y (2023) Federated learning-based cross-enterprise recommendation with graph neural networks. IEEE Trans Industr Inform 19(1):673–682
29. Li Z, Li Q, Zhou Y, Zhong W, Zhang G, Wu C (2023) Edge-cloud collaborative learning with federated and centralized features. In Proceedings of the 46th International ACM SIGIR conference on research and development in information retrieval, SIGIR '23. Association for Computing Machinery, New York, pp 1949–1953
30. Li Z, Lin Z, Liang F, Pan W, Yang Q, Ming Z (2024) Decentralized federated recommendation with privacy-aware structured client-level graph. ACM Trans Intell Syst Technol 15(4):1–23
31. Li Z, Liu F, Yang W, Peng S, Zhou J (2022) A survey of convolutional neural networks: analysis, applications, and prospects. IEEE Trans Neural Networks Learn Syst 33(12): 6999–7019
32. Liang F, Pan W, Ming Z (2021) Fedrec++: Lossless federated recommendation with explicit feedback. In Proceedings of the AAAI conference on artificial intelligence
33. Lin G, Liang F, Pan W, Ming Z (2021) Fedrec: Federated recommendation with explicit feedback. IEEE Intell Syst 36(5):21–30
34. Lin Y, Ren P, Chen Z, Ren Z, Yu D, Ma J, Rijke Md, Cheng X (2020) Meta matrix factorization for federated rating predictions. In Proceedings of the 43rd International ACM SIGIR conference on research and development in information retrieval. ACM Press, New York, pp 981–990
35. Lin Z, Pan W, Yang Q, Ming Z (2022) A generic federated recommendation framework via fake marks and secret sharing. ACM Trans Inf Syst 41(2):1–37
36. Liu Z, Chen Y, Xia F, Bian J, Zhu B, Shen G, Kong X (2023) Tap: traffic accident profiling via multi-task spatio-temporal graph representation learning. ACM Trans Knowl Discov Data 17(4):1–25
37. Liu Z, Song L, Fragouli C (2022) Federated multi-armed bandits with vector rewards for aspect-based recommendations. In IEEE Global Communications Conference, pp 1079–1084

38. Liu Z, Yang L, Fan Z, Peng H, Yu PS (2022) Federated social recommendation with graph neural network. ACM Trans Intell Syst Technol 13(4):1–24
39. Luo L, Liu B (2022) Dual-contrastive for federated social recommendation. In International Joint Conference on Neural Networks
40. Luo S, Xiao Y, Song L (2022) Personalized federated recommendation via joint representation learning, user clustering, and model adaptation. In Proceedings of the 31st ACM international conference on information & knowledge management, CIKM '22. Association for Computing Machinery, New York, pp 4289–4293
41. Luo S, Xiao Y, Zhang X, Liu Y, Ding W, Song L (2024) Perfedrec++: enhancing personalized federated recommendation with self-supervised pre-training. ACM Trans Intell Syst Technol 15(5):1–24
42. Ma C, Ren X, Xu G, He B (2023) FedGR: federated graph neural network for recommendation systems. Axioms 12(2):170
43. McMahan B, Moore E, Ramage D, Hampson S, Arcas BAy (2017) Communication-efficient learning of deep networks from decentralized data. In Proceedings of the 20th international conference on artificial intelligence and statistics. PMLR, New York
44. Minto L, Haller M, Livshits B, Haddadi H (2021) Stronger privacy for federated collaborative filtering with implicit feedback. In Fifteenth ACM Conference on Recommender Systems. ACM Press, New York, pp 342–350
45. Muhammad K, Wang Q, O'Reilly-Morgan D, Tragos E, Smyth B, Hurley N, Geraci J, Lawlor A (2020) Fedfast: going beyond average for faster training of federated recommender systems. In Proceedings of the 26th ACM SIGKDD international conference on knowledge discovery & data mining, KDD '20. ACM Press, New York, pp 1234–1242
46. Nichol A, Achiam J, Schulman J (2018) On first-order meta-learning algorithms. arXiv:1803.02999
47. Ning L, Singhal K, Zhou EX, Prakash S (2021) Learning federated representations and recommendations with limited negatives. arXiv:2108.07931
48. Pateria S, Subagdja B, Tan Ah, Quek C (2021) Hierarchical reinforcement learning: a comprehensive survey. ACM Comput Surv 54(5):1–35
49. Peng Z, Yang Y, Yang L, Chen K (2022) Federated meta embedding concept stock recommendation. IEEE Trans Big Data
50. Perifanis V, Drosatos G, Stamatelatos G, Efraimidis PS (2023) FedPOIRec: privacy-preserving federated poi recommendation with social influence. Inf Sci 623:767–790
51. Perifanis V, Efraimidis PS (2022) Federated neural collaborative filtering. Knowl-Based Syst 242:108441
52. Pouyanfar S, Sadiq S, Yan Y, Tian H, Tao Y, Reyes MP, Shyu ML, Chen SC, Iyengar SS (2018) A survey on deep learning: Algorithms, techniques, and applications. ACM Comput Surv 51(5):1–36
53. Qi T, Wu F, Wu C, Huang Y, Xie X (2020) Privacy-preserving news recommendation model learning. In Findings of the Association for Computational Linguistics: EMNLP 2020. Association for Computational Linguistics, New York, pp 1423–1432
54. Qu L, Tang N, Zheng R, Nguyen QVH, Huang Z, Shi Y, Yin H (2023) Semi-decentralized federated ego graph learning for recommendation. In Proceedings of the ACM Web Conference 2023, WWW '23. Association for Computing Machinery, New York, pp 339–348
55. Rendle S, Freudenthaler C, Gantner Z, Schmidt-Thieme L (2009) BPR: Bayesian personalized ranking from implicit feedback. In Proceedings of the twenty-fifth conference on uncertainty in artificial intelligence, UAI '09. AUAI Press, Arlington, pp 452–461
56. Rifai S, Vincent P, Muller X, Glorot X, Bengio Y (2011) Contractive auto-encoders: Explicit invariance during feature extraction. In International conference on machine learning. https://api.semanticscholar.org/CorpusID:8141422
57. Ryu S, Takamaeda-Yamazaki S (2022) Model-based federated reinforcement distillation. In IEEE Global Communications Conference, pp 1109–1114
58. Salakhutdinov R, Mnih A (2007) Probabilistic matrix factorization. In Proceedings of the 20th international conference on neural information processing systems, NIPS'07. Curran Associates Inc., Red Hook, pp 1257–1264

59. Shao H, Xiao Z, Yao S, Sun D, Zhang A, Liu S, Wang T, Li J, Abdelzaher T (2022) Controlvae: tuning, analytical properties, and performance analysis. IEEE Trans Pattern Anal Mach Intell 44(12):9285–9297
60. Shi C, Shen C (2021) Federated multi-armed bandits. In Proceedings of the AAAI conference on artificial intelligence, pp 9603–9611
61. Shi Y, Huang Z, Feng S, Zhong H, Wang W, Sun Y (2021) Masked label prediction: unified message passing model for semi-supervised classification. In International joint conferences on artificial intelligence
62. Singh AP, Gordon GJ (2008) Relational learning via collective matrix factorization. In Proceedings of the 14th ACM SIGKDD international conference on knowledge discovery and data mining, KDD '08. ACM Press, New York, pp 650–658
63. Tan AZ, Yu H, Cui L, Yang Q (2023) Towards personalized federated learning. IEEE Trans Neural Networks Learn Syst 34(12):9587–9603
64. Tang J, Wang K (2018) Personalized top-n sequential recommendation via convolutional sequence embedding. In Proceedings of the Eleventh ACM international conference on web search and data mining, WSDM '18. Association for Computing Machinery, New York, , pp 565–573
65. Thapa C, Arachchige PCM, Camtepe S, Sun L (2022) Splitfed: when federated learning meets split learning. In Proceedings of the AAAI conference on artificial intelligence, pp 8485–8493
66. Tommasel A, Rodriguez JM, Godoy D (2021) I want to break free! recommending friends from outside the echo chamber. In Fifteenth ACM Conference on Recommender Systems. ACM Press, New York
67. Vincent P, Larochelle H, Bengio Y, Manzagol PA (2008) Extracting and composing robust features with denoising autoencoders. In Proceedings of the 25th international conference on machine learning, ICML '08. Association for Computing Machinery, New York. https://doi.org/10.1145/1390156.1390294
68. Wahab OA, Rjoub G, Bentahar J, Cohen R (2022) Federated against the cold: a trust-based federated learning approach to counter the cold start problem in recommendation systems. Inf Sci 601:189–206
69. Wang J, Zhang T, Song J, Sebe N, Shen HT (2018) A survey on learning to hash. IEEE Trans Pattern Anal Mach Intell 40(4):769–790
70. Wang Q, Yin H, Chen T, Yu J, Zhou A, Zhang X (2021) Fast-adapting and privacy-preserving federated recommender system. VLDB J 31(5):877–896
71. Wei S, Meng S, Li Q, Zhou X, Qi L, Xu X (2023) Edge-enabled federated sequential recommendation with knowledge-aware transformer. Futur Gener Comput Syst 148:610–622
72. Wu C, Wu F, Lyu L, Qi T, Huang Y, Xie X (2022) A federated graph neural network framework for privacy-preserving personalization. Nat. Commun. 13(1):3091
73. Wu J, Liu Q, Huang Z, Ning Y, Wang H, Chen E, Yi J, Zhou B (2021) Hierarchical personalized federated learning for user modeling. In Proceedings of the web conference, WWW '21. ACM Press, New York, pp 957–968
74. Xia F, Wang L, Tang T, Chen X, Kong X, Oatley G, King I (2023) Cengcn: centralized convolutional networks with vertex imbalance for scale-free graphs. IEEE Trans Knowl Data Eng 35(5):4555–4569
75. Yan B, Cao Y, Wang H, Yang W, Du J, Shi C (2024) Federated heterogeneous graph neural network for privacy-preserving recommendation. In Proceedings of the Web Conference
76. Yang E, Huang Y, Liang F, Pan W, Ming Z (2021) FCMF: federated collective matrix factorization for heterogeneous collaborative filtering. Knowl-Based Syst 220:106946
77. Yang L, Zhang J, Chai D, Wang L, Guo K, Chen K, Yang Q (2022) Practical and secure federated recommendation with personalized mask. In International workshop on trustworthy federated learning, pp 33–45
78. Ye Z, Zhang X, Chen X, Xiong H, Yu D (2023) Adaptive clustering based personalized federated learning framework for next poi recommendation with location noise. IEEE Trans Knowl Data Eng 36(5):1843–1856

79. Yi J, Wu F, Wu C, Liu R, Sun G, Xie X (2021) Efficient-FedRec: efficient federated learning framework for privacy-preserving news recommendation. arXiv preprint arXiv:2109.05446
80. Yin Y, Li Y, Gao H, Liang T, Pan Q (2023) FGC: GCN-based federated learning approach for trust industrial service recommendation. IEEE Trans Industr Inform 19(3):3240–3250
81. Zhang C, Long G, Zhou T, Yan P, Zhang Z, Yang B (2023) Graph-guided personalization for federated recommendation. arXiv-2305
82. Zhang C, Long G, Zhou T, Yan P, Zhang Z, Zhang C, Yang B (2023) Dual personalization on federated recommendation. In International Joint Conferences on Artificial Intelligence
83. Zhang G, Liu Y, Jin X (2020) A survey of autoencoder-based recommender systems. Front Comp Sci 14:430-450
84. Zhang H, Luo F, Wu J, He X, Li Y (2023) LightFR: Lightweight federated recommendation with privacy-preserving matrix factorization. ACM Trans Inf Syst 41(4):1–28
85. Zhang K, Sun T, Tao Y, Genc S, Mallya S, et al (2020) Robust multi-agent reinforcement learning with model uncertainty. Adv Neural Inf Proces Syst 33:10571–10583
86. Zhang R, Chen Y, Wu C, Wang F (2023) Cluster-driven gnn-based federated recommendation with biased message dropout. In IEEE International Conference on Multimedia and Expo, pp 594–599
87. Zhang X, Ye Z, Lu J, Zhuang F, Zheng Y, Yu D (2023) Fine-grained preference-aware personalized federated poi recommendation with data sparsity. In Proceedings of the 46th International ACM SIGIR conference on research and development in information retrieval, SIGIR '23. Association for Computing Machinery, New York, pp 413–422
88. Zhao S, Bharati R, Borcea C, Chen Y (2020) Privacy-aware federated learning for page recommendation. In IEEE international conference on big data, pp 1071–1080
89. Zheng X, Wang Z, Chen C, Qian J, Yang Y (2023) Decentralized graph neural network for privacy-preserving recommendation. In Proceedings of the 32nd ACM international conference on information and knowledge management, CIKM '23. Association for Computing Machinery, New York, pp 3494–s3504

Chapter 3
Privacy Computing in Cross-Device Federated Recommendation

Abstract This chapter explores critical aspects of privacy computing in federated recommendation systems. It provides an overview of homomorphic encryption, discussing partial and full homomorphism alongside their advantages and limitations. Key techniques in differential privacy, such as additive noise and randomized response, are evaluated for effectiveness. Secure multi-party computation is discussed with methods like Secret Sharing (SS) and Garbled Circuit. The chapter also addresses attack vectors, including poisoning and inference attacks, and strategies for Byzantine resilience. This analysis aims to strengthen privacy-preserving techniques in federated learning environments.

3.1 Homomorphic Encryption

RecSys rely heavily on users' preferences to generate accurate and personalized recommendations. However, even after removing explicit identifiers like names and phone numbers, raw preference data can still expose sensitive information about users. For instance, Narayanan et al. [41] demonstrated that specific subscribers' records in the Netflix Prize dataset could be reidentified using background knowledge from the Internet Movie Database (IMDb). Similarly, Weinsberg et al. [54] were able to infer users' gender based on their ratings in the Flixster and MovieLens datasets. In the context of point-of-interest (POI) recommendation, check-in data could reveal sensitive information about users, such as their identity, health status, or income level [63, 68]. These studies underscore the importance of privacy-preserving techniques in RecSys.

Homomorphic encryption (HE) is a cryptographic technique that encrypts plaintexts using keys and allows computations to be performed on ciphertexts without the need for decryption [2]. This unique feature of HE ensures data privacy and security even during the computation process. By allowing computations to be conducted on encrypted data, HE effectively eliminates the risk of data exposure, even if an adversary gains access to the data during processing. This capability is particularly valuable in distributed systems like federated learning, where data is often processed

X. Kong et al., *Cross-device Federated Recommendation*, Machine Learning: Foundations, Methodologies, and Applications,
https://doi.org/10.1007/978-981-96-3212-1_3

across multiple decentralized nodes. Generally, HE can be classified along three dimensions:

1. **Symmetry and Asymmetry**:
   - **Symmetric Encryption**: This type of encryption utilizes the same key for both encryption and decryption processes. It is known for being simpler and faster compared to asymmetric encryption. However, in federated systems, symmetric encryption presents significant security challenges because the secret key must be shared among multiple parties, increasing the risk of key exposure. The challenge of securely distributing the key among participants without compromising it adds a layer of complexity that can be difficult to manage, particularly in large-scale systems.
   - **Asymmetric Encryption**: This approach uses a pair of keys—a public key for encryption and a private key for decryption. Although asymmetric encryption is computationally more demanding, it offers enhanced security suitable for distributed environments, where maintaining data privacy and security is paramount. The public-private key pair mechanism ensures that even if the public key is shared, the data remains secure because only the corresponding private key can decrypt the information. This makes it ideal for scenarios where data needs to be shared among multiple parties without risking unauthorized access.

2. **Addition, Multiplication, and Hybrid**:
   - **Additive Homomorphism**: This form of HE supports addition operations on ciphertexts. It is particularly beneficial for applications requiring data aggregation, such as summing user ratings or calculating average preferences. Additive homomorphism simplifies the process of combining encrypted data points without exposing individual values. This capability is crucial in federated learning, where the aggregation of encrypted data from multiple sources is often required to train a global model.
   - **Multiplicative Homomorphism**: This type of homomorphic encryption supports multiplication operations on encrypted data, making it particularly useful for complex computations like polynomial evaluations or weighted sums, which are essential in advanced machine learning algorithms. In federated learning, where complex model updates are often required, such as in training processes that involve gradient descent, this capability allows these updates to be performed on encrypted data, thus ensuring the privacy of individual contributions while still enabling effective model training.
   - **Hybrid Homomorphism**: Combining both additive and multiplicative operations, hybrid homomorphism provides a versatile solution that accommodates a broad range of computational tasks while maintaining data security. This flexibility is critical for complex machine learning algorithms that require both addition and multiplication operations. The hybrid approach is particularly advantageous in scenarios where different types of operations are needed to achieve the desired computational outcome, making it a robust choice for federated learning applications.

3. **Partially, Somewhat, and Fully HE**:
   - **Partially Homomorphic Encryption (PHE)**: PHE supports an unlimited number of a single type of operation (either addition or multiplication). This makes it suitable for applications requiring simple, repetitive tasks where computational efficiency is prioritized. PHE's efficiency is particularly beneficial in scenarios where a large number of similar operations need to be performed on encrypted data, such as in basic statistical calculations or data aggregation tasks.
   - **Somewhat Homomorphic Encryption (SHE)**: SHE supports a limited number of both addition and multiplication operations, offering a balance between functionality and computational efficiency. It allows for a predefined number of operations, sufficient for many practical applications where both types of operations are needed but not excessively. SHE is well-suited for scenarios in federated learning where a moderate level of computational complexity is required without overwhelming the system's resources.
   - **Fully Homomorphic Encryption (FHE)**: FHE provides the most comprehensive capabilities, supporting an unlimited number of both addition and multiplication operations. This makes FHE ideal for complex, long-term computations in federated learning environments, despite its higher computational overhead. FHE is crucial for scenarios that demand high versatility and robust security, as it enables any computation to be performed on encrypted data without ever decrypting it, thus ensuring the highest level of privacy and security.

In the following subsections, we categorize related works into partial and full homomorphism. These categorizations help in understanding the specific applications and limitations of different HE schemes in practical federated learning scenarios. Partial homomorphism is often employed where computational efficiency is paramount, while full homomorphism is favored in scenarios that require extensive functionality and stringent security measures. Continuous advancements in HE technology are addressing these trade-offs, making the implementation of HE in real-world federated systems increasingly feasible. The ongoing development in this field is driving the adoption of HE in more complex and varied federated learning applications, where the need for both security and computational efficiency is paramount.

The use of HE in federated learning environments ensures that data remains encrypted throughout the computational process, significantly enhancing privacy and security. By allowing operations on encrypted data, HE minimizes the risk of data breaches and unauthorized access, thus fostering trust and encouraging data sharing among participants. This trust is essential for the successful deployment and adoption of federated learning systems, which rely on the collaborative efforts of multiple data owners. Moreover, the ability of HE to maintain data privacy without compromising the accuracy and effectiveness of machine learning models is a key factor in its growing popularity. As the field continues to evolve, HE is expected to play an increasingly important role in the development of secure, privacy-preserving federated learning systems.

3.1.1 Partial Homomorphism

From our observation, current studies on HFedRec mainly exploit Paillier HE [42] to protect transmitted gradients or model weights. Paillier HE is an *asymmetric*, *additive*, and *partial* encryption schema. This means it uses a public-private key pair for encryption and decryption, supports addition operations on encrypted data, and is limited to specific types of computations. The choice of Paillier HE in HFedRec systems is driven by its ability to perform homomorphic addition, which allows the secure aggregation of encrypted data without needing decryption, thus preserving the privacy of the underlying data. Based on Paillier HE, FedMF [11] first defines the encrypted training steps for secure matrix factorization, which is a fundamental process in recommendation systems:

- **Step 1.** KeyGen $\rightarrow (pk, sk)$: An honest client generates a public-private key pair of (pk, sk), where pk is distributed to all parties and sk is only held by clients. This key generation step is crucial in ensuring that the encrypted data can be securely shared across different entities involved in the federated learning process while keeping the decryption key confidential.
- **Step 2.** Enc$(\boldsymbol{V}, pk) \rightarrow C_{\boldsymbol{V}}$: The central server encrypts the initialized item embedding matrix $\boldsymbol{V}$ with pk to obtain $C_{\boldsymbol{V}}$, and it is available for download by all clients. This step ensures that the item embeddings are protected from potential eavesdroppers or malicious entities during transmission.
- **Step 3.** Dec$(C_{\boldsymbol{V}}, sk) \rightarrow \boldsymbol{V}$: Each client downloads the latest encrypted matrix $C_{\boldsymbol{V}}$ from the central server and deciphers it with the sk to obtain $\boldsymbol{V}$. The decryption process at the client side allows them to work with the actual data while ensuring that the transmitted data remains encrypted and secure.
- **Step 4.** Clients locally update the user embedding and calculate items' gradients ∇_g. This local computation step is fundamental in federated learning, where clients perform model updates based on their private data without sharing it directly with the central server.
- **Step 5.** Enc$(\nabla_g, pk) \rightarrow C_{\nabla_g}$: Items' gradients are encrypted with pk and sent to the server. The encryption of gradients ensures that the model updates transmitted back to the server do not expose any sensitive information about the users' data.
- **Step 6.** The central server can aggregate submitted gradients C_{∇_g} and update global item embeddings $C_{\boldsymbol{V}}$ via homomorphic properties:

$$\begin{aligned} \mathrm{D}(\mathrm{E}(m_i) \times \mathrm{E}(m_j)) &= m_i + m_j \\ \mathrm{D}(\mathrm{E}(m)^k (\bmod n^2)) &= km (\bmod n) \end{aligned} \tag{3.1}$$

where E(·) and D(·) are encryption and decryption algorithms and m_i and m_j denote plaintexts. This aggregation step leverages the additive homomorphic properties of Paillier HE to securely combine the encrypted gradients without decryption, thus preserving data privacy during the update process.

- **Step 7.** Repeat steps 3–6 until convergence. This iterative process continues until the model converges, ensuring that the updates are securely aggregated and the model is trained without compromising data privacy.

Following variants of federated MF [4, 26] adopt a similar procedure to protect data transmission. These steps can be easily extended to deep learning-based HFedRec [60] when the centrally aggregated data are embeddings or hidden layers' weights. The encryption and aggregation steps ensure that the data remains secure during transmission and processing, preventing unauthorized access and maintaining the privacy of individual users' data. This approach demonstrates the versatility of Paillier HE in various machine learning contexts, particularly in federated learning where privacy and security are critical concerns.

Differently, FedPerGNN [18] designs a local graph expansion schema to incorporate high-order connectivity, and keys are generated in the server. The schema introduces a third party to collect user embeddings and encrypted item IDs. Through matching item sequences, the third party can provide each user with user embeddings of her neighbors who have similar behaviors. This approach not only enhances the privacy of the system by adding an additional layer of security but also improves the recommendation accuracy by considering the connectivity and relationships between different users. By leveraging the structure of the user-item interaction graph, FedPerGNN can achieve more accurate recommendations while preserving user privacy through the use of encrypted data and secure computation techniques.

Moreover, the use of Paillier HE in HFedRec systems has demonstrated significant potential in enhancing both security and performance. By leveraging the additive homomorphic properties, these systems can perform secure computations on encrypted data, enabling privacy-preserving training and inference processes. This makes Paillier HE an ideal choice for federated learning scenarios where data privacy and security are of utmost importance. The ability to securely aggregate encrypted data without decrypting it ensures that sensitive information remains protected throughout the entire learning process, thereby addressing one of the primary challenges in federated learning.

In conclusion, the implementation of partial homomorphic encryption, particularly Paillier HE, in HFedRec systems represents a significant advancement in privacy-preserving machine learning. By enabling secure aggregation and computation of encrypted data, it addresses critical privacy concerns while maintaining the efficiency and accuracy of the recommendation algorithms; Fig. 3.1 shows the application of the Paillier HE algorithm. As research in this area continues to evolve, we can expect further improvements in the scalability and robustness of these systems, paving the way for more widespread adoption in various federated learning applications. The ongoing advancements in partial homomorphic encryp-

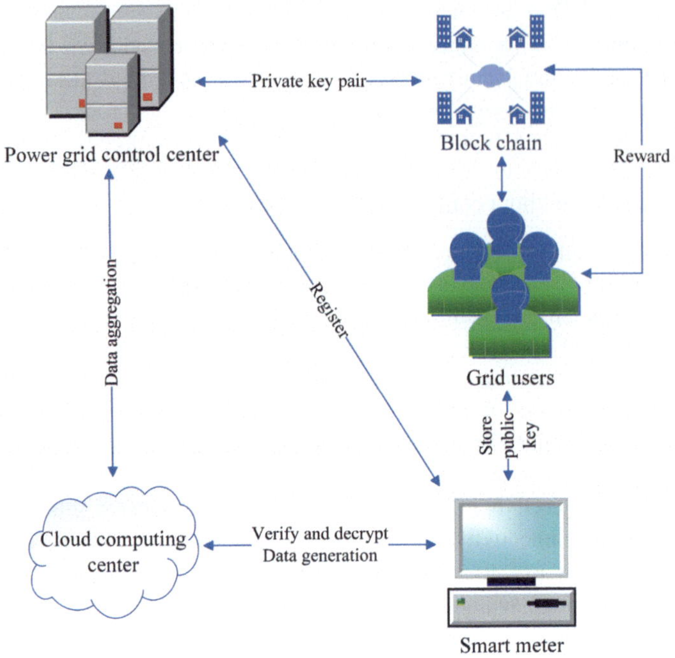

Fig. 3.1 Application of Paillier HE algorithm

tion techniques, coupled with their successful application in HFedRec, suggest a promising future for secure and privacy-preserving federated learning frameworks.

3.1.2 Full Homomorphism

Paillier HE only allows calculations of positive integers, and thus extra transformations are required to process float and negative numbers. To facilitate computational flexibility, FedPOIRec [30] utilizes Cheon-Kim-Kim-Song (CKKS) HE [14] to search nearest user embeddings as social neighbors. The social integration protocol is a deep promotion of FedPerGNN's and formulated as four detailed phases. As an *asymmetric*, *hybrid*, and *full* encryption schema, CKKS HE provides the protocol with additive and multiplicative operations on real numbers, making it practical to compute cosine similarities between user embeddings; Fig. 3.2 shows the application of additive homomorphism in a federated environment.

The CKKS homomorphic encryption scheme represents a significant advancement in the field of cryptography and secure computation. Its design, optimized for approximate arithmetic on real numbers, bridges a critical gap between theoretical cryptographic constructs and practical applications in data science and machine learning. CKKS HE allows for the encryption of floating-point numbers

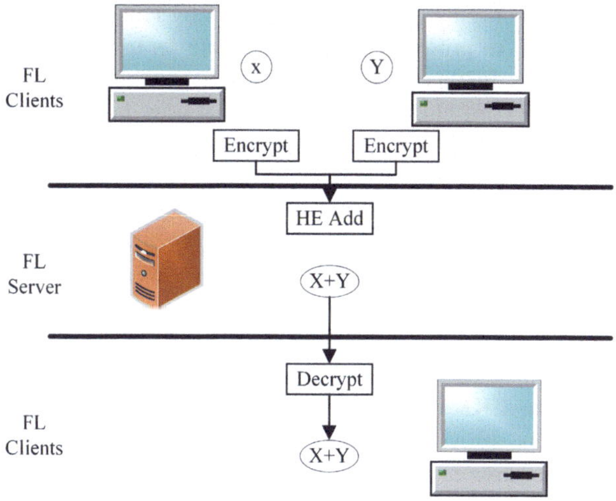

Fig. 3.2 Homomorphic encryption

and supports both addition and multiplication operations on ciphertexts, which is essential for many machine learning algorithms that require these types of operations. By enabling these operations on encrypted data, CKKS HE ensures that sensitive information remains confidential throughout the computation process, thus providing a strong foundation for privacy-preserving machine learning in federated environments.

CKKS achieves this by encoding real numbers into complex numbers and packing multiple plaintext slots into a single ciphertext. This packing technique significantly increases the efficiency of homomorphic operations, making CKKS particularly well-suited for applications in federated learning where large-scale data processing is required. The scheme supports rescaling, which maintains the precision of the computations by adjusting the scale of the ciphertexts after multiplication. This is crucial for ensuring that the results of encrypted computations remain accurate and useful. The ability to perform precise arithmetic operations on encrypted data without significant loss of accuracy is one of the key features that make CKKS HE a powerful tool for secure machine learning applications.

As federated learning continues to gain traction as a paradigm for privacy-preserving machine learning, the role of CKKS in enabling secure and efficient encrypted computations cannot be overstated. This synergy between CKKS and federated learning not only enhances data security but also opens up new frontiers for collaborative and confidential data analysis. By enabling secure computation of cosine similarities, CKKS HE facilitates the identification of nearest neighbors in a privacy-preserving manner, which is a fundamental operation in many recommendation systems. This capability is particularly important in scenarios where the preservation of user privacy is paramount, yet accurate and personalized recommendations are still required.

Beyond securely averaging uploaded parameters by Paillier HE, FINDING [70] also employs fully HE to protect k-means clustering on user representations from privacy disclosure [9]. In this context, CKKS HE is used to perform encrypted computations necessary for clustering, ensuring that user data remains confidential throughout the process. The use of fully homomorphic encryption allows for the development of cluster-level models that are personalized for federated learning without exposing individual user data. This method ensures that the clustering process, which is critical for identifying similar user groups, can be performed securely and efficiently, maintaining the privacy of the users involved.

The incorporation of CKKS HE into these systems illustrates the practical application of advanced cryptographic techniques to enhance privacy and security in federated learning. As research progresses, it is expected that further optimizations and improvements in CKKS HE will continue to drive its adoption in a wider range of applications, providing robust solutions for secure and efficient data processing in collaborative environments. The ongoing advancements in CKKS HE, including improvements in computational efficiency and scalability, suggest that this encryption scheme will play an increasingly important role in the future of privacy-preserving machine learning.

In practical terms, the use of CKKS HE in FedPOIRec allows for the calculation of cosine similarities between encrypted user embeddings, which is a crucial operation for identifying social neighbors. The protocol involves encrypting user embeddings, performing the similarity calculations on the encrypted data, and then decrypting the results, ensuring that sensitive user information is never exposed during the process. This method significantly enhances the privacy of the recommendation system while maintaining its effectiveness. By protecting the integrity and confidentiality of user data throughout the computation process, CKKS HE enables the development of secure and trustworthy recommendation systems.

FINDING demonstrates another application of CKKS HE by employing it to protect k-means clustering on user representations. This involves encrypting user data before clustering operations and performing all necessary calculations in the encrypted domain. By doing so, FINDING ensures that individual user data remains confidential and secure throughout the clustering process, addressing critical privacy concerns in federated learning environments. The ability to perform complex clustering operations on encrypted data without compromising privacy is a significant advantage in scenarios where user data must remain secure.

These examples highlight the versatility and effectiveness of CKKS HE in providing full homomorphic encryption capabilities. The ability to perform complex computations on encrypted data without compromising privacy is a significant advantage in federated learning. As CKKS HE continues to evolve, its application in privacy-preserving machine learning is expected to expand, offering more robust and scalable solutions for secure data processing in various collaborative settings. The growing interest in CKKS HE within the research community suggests that it will become an increasingly important tool for ensuring data security and privacy in a wide range of applications.

In conclusion, the implementation of full homomorphic encryption, particularly CKKS HE, in HFedRec systems represents a critical advancement in ensuring data privacy and security. By enabling secure computations on encrypted data, CKKS HE addresses significant privacy concerns while maintaining the efficiency and accuracy of recommendation algorithms. The ongoing research and development in this field promise further enhancements in the scalability and robustness of these systems, paving the way for broader adoption in diverse federated learning applications. The continued innovation in CKKS HE and its application to federated learning will likely drive the development of more secure, efficient, and privacy-preserving machine learning frameworks in the coming years.

3.1.3 Summary and Challenges

Though the encryption-decryption process may accumulate noise, the error fluctuation range is acceptable as evaluated in [11, 26]. Hence, HE guarantees both accuracy requirements and secure parameter exchange for HFedRec. Homomorphic encryption (HE) has demonstrated its efficacy in ensuring privacy and security in federated recommendation systems, providing a robust framework for secure computations on encrypted data. By enabling operations on ciphertexts, HE eliminates the need to decrypt data, thus preserving the confidentiality of sensitive information throughout the computational process.

Nevertheless, the key pair generation and frequent encryption-decryption processes may lead to intolerable storage requirements, computational complexity, and communication overhead [11, 30, 60]. The high computational demands associated with HE, particularly in generating and managing key pairs, can be a significant burden for local devices with limited processing power and storage capacity. This challenge is exacerbated in federated learning environments where numerous devices must frequently encrypt and decrypt data, leading to potential performance bottlenecks.

Owing to the limited power of local devices, the intrinsic challenge is how to trade off security and efficiency, especially in the cross-user federation. Balancing the rigorous security requirements with the need for efficient computation and communication remains a critical issue. As the server solely performs the parameter aggregation, some homomorphic properties are redundant and can be refined for efficiency enhancement [55]. For instance, optimizing the use of homomorphic properties by reducing unnecessary operations can significantly enhance computational efficiency without compromising security. This involves identifying and eliminating redundant homomorphic operations that do not contribute to the overall security of the system.

Additionally, advancements in cryptographic techniques and computational optimizations are essential to address these challenges. Exploring lightweight HE schemes or hybrid approaches that combine HE with other privacy-preserving techniques can potentially reduce the computational and storage burdens. For example, integrating differential privacy with homomorphic encryption could help in achieving a balance between privacy and efficiency, as differential privacy can be used to add noise to the data before encryption, reducing the computational load during encryption and decryption phases. Another avenue is the development of more efficient algorithms for key generation and management, as well as improved methods for ciphertext compression to mitigate storage overhead. These advancements could involve the use of elliptic curve cryptography (ECC) for more efficient key generation or the adoption of lattice-based cryptography for enhanced security with lower computational costs.

Moreover, it is crucial to investigate strategies for optimizing communication protocols to reduce the bandwidth required for transmitting encrypted data. This includes developing more efficient encryption schemes that produce smaller ciphertexts and leveraging advanced compression techniques to minimize the size of encrypted data packets. One promising direction is the use of adaptive compression techniques that adjust the compression level based on the available bandwidth and the specific requirements of the federated learning task. Furthermore, optimizing the network infrastructure to support more efficient data transmission, such as implementing edge computing solutions to process data closer to the source, could significantly alleviate the communication burden in federated learning scenarios.

The future of HE in federated learning will likely involve a combination of these approaches, focusing on enhancing both security and efficiency. By addressing the inherent trade-offs and pushing the boundaries of current cryptographic research, it is possible to develop more practical and scalable solutions for privacy-preserving federated learning. Continued collaboration between cryptographers, data scientists, and system engineers will be key to overcoming these challenges and realizing the full potential of HE in real-world applications. For instance, interdisciplinary efforts could lead to the creation of standardized HE frameworks that are specifically designed for federated learning environments, facilitating their broader adoption and integration into existing systems.

The goal is to create a federated learning environment where data privacy is rigorously protected without sacrificing performance. This involves not only optimizing the technical aspects of HE but also considering the usability and accessibility of these techniques for various stakeholders, including developers, end users, and organizations. As the field evolves, the integration of more sophisticated HE techniques and optimizations will play a crucial role in achieving this balance, paving the way for broader adoption of secure and efficient federated learning systems. Achieving this will require not only technical advancements but also standardized protocols and frameworks that facilitate the widespread implementation of HE in diverse federated learning applications.

In conclusion, while HE offers a powerful solution for privacy-preserving federated learning, addressing the challenges of computational complexity, storage requirements, and communication overhead is essential for its practical adoption. By focusing on optimizing and refining these aspects, the federated learning community can ensure that HE remains a viable and effective tool for securing sensitive data in distributed machine learning environments. The ongoing research and development in this field will be pivotal in shaping the future of secure, efficient, and scalable federated learning systems. As more real-world applications of federated learning emerge, the need for robust and efficient HE solutions will only grow, underscoring the importance of continued innovation and collaboration in this area.

3.2 Differential Privacy

Differential privacy (DP) aims to add extra noise and prevent computational outputs from being sensitive to the presence or absence of a specific data point [16]. This technique is crucial in ensuring that individual data points do not significantly influence the results of data analysis, thereby preserving the privacy of users. The fundamental goal of DP is to provide a framework that allows data analysts to extract useful insights from datasets while minimizing the risk of exposing sensitive information about individuals. There are different models of DP, including central DP (CDP) and local DP (LDP), each offering unique approaches to balancing privacy and data utility.

In central DP (CDP), a trusted third party is required to store data centrally and respond to external query requests with statistical information that does not reveal individual characteristics. Given privacy loss parameters (ϵ, δ) and a randomized algorithm M, for any two neighboring datasets X and X' (differing in a single entry $x \in X$, $X' = X\backslash\{x\}$) and output $O \in \text{Range(M)}$, (ϵ, δ)-differential privacy is satisfied when

$$\Pr[\mathrm{M}(X) \in O] \leq e^{\epsilon}\Pr[\mathrm{M}(X') \in O] + \delta \tag{3.2}$$

where a greater privacy budget ϵ means a greater difference between the probability distributions of $\mathrm{M}(X)$ and $\mathrm{M}(X')$. The definition guarantees that $\mathrm{M}(X)$ and $\mathrm{M}(X')$ have similar probability distributions but permit a small deviation range δ. This ensures that the inclusion or exclusion of a single data point does not significantly affect the outcome, thus providing privacy protection for individual data points. CDP relies on the centralization of data, which, while effective in traditional data processing environments, poses challenges when applied to decentralized systems like federated learning.

Local DP (LDP) does not require a trusted third party and inherits the ability to quantitatively analyze privacy-exposure degrees from CDP, such as post-processing and composition theorems [16]. Despite introducing perturbation to each user and increasing the total amount of noise, LDP accomplishes fine-grained privatization of original data [67]. Each user perturbs their own data before sharing it, which enhances privacy but may also lead to higher overall noise levels in the aggregated data; Fig. 3.3 shows the application of user-level differential privacy. The decentralized nature of LDP makes it particularly suitable for settings where users are unwilling or unable to trust a central entity with their data. By ensuring that the data

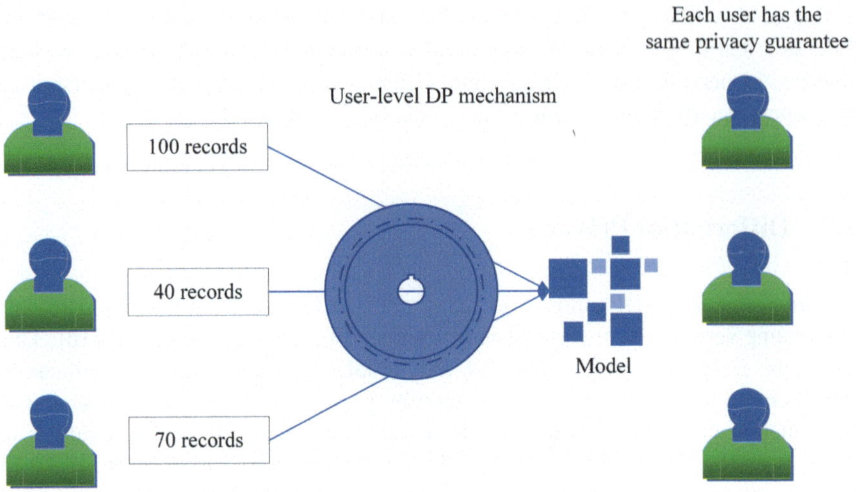

Fig. 3.3 User-level differential privacy

is perturbed before it leaves the user's device, LDP offers strong privacy guarantees even in adversarial environments.

In the context of federated learning, raw data is retained locally on users' devices, and we do not have visibility into the differences between users' datasets at the granularity of a single sample. This makes the direct application of CDP challenging in FL settings. However, the principles of DP can still be applied to ensure privacy in FL. The challenge in federated learning lies in balancing the need for data privacy with the requirement for accurate and useful model updates. DP offers a way to achieve this balance by introducing carefully calibrated noise into the learning process, thus obscuring the contributions of individual data points without significantly degrading the overall performance of the model.

For instance, DP can be integrated into the federated learning process by adding noise to the model updates rather than the raw data. This approach, known as federated DP, helps to preserve the privacy of the individual data points while allowing the model to learn from the aggregated data. The noise added to the model updates ensures that the contribution of any single data point is obfuscated, thus providing privacy guarantees. This method allows federated learning systems to benefit from the privacy protections offered by DP while still enabling effective collaboration among distributed participants.

The effectiveness of DP in federated learning can be further enhanced by employing techniques such as adaptive noise scaling, where the amount of noise added to the model updates is adjusted based on the sensitivity of the data or the specific phase of the learning process. This approach helps to maintain a balance between privacy and utility, ensuring that the model remains accurate while still providing strong privacy guarantees.

The following subsections will depict specific DP techniques in the federated setting, illustrating how these methods can be effectively employed to balance privacy and utility. These techniques will explore the implementation of both CDP and LDP principles within the federated learning framework, addressing the unique challenges posed by decentralized data and ensuring robust privacy protection for users. By applying these techniques, federated learning systems can achieve a high level of privacy without compromising on the quality of the insights generated from the data. As the field of DP continues to evolve, we can expect to see more sophisticated methods that further enhance the privacy guarantees offered by federated learning systems, making them increasingly viable for a wide range of applications where data privacy is a critical concern.

3.2.1 Additive Noise

Additive noise techniques involve introducing noise from an independent distribution to the query function f to mask the true gradients submitted to the server. The basic formulation is the following:

$$\mathrm{M}(\nabla_g) := f(\nabla_g) + \mathrm{Noise}(\alpha, \beta) \tag{3.3}$$

where ∇_g represents the gradients submitted to the server for aggregation, and $\mathrm{Noise}(\cdot)$ can be a Laplace or Gaussian distribution. This method helps in ensuring that the presence or absence of any individual user in the dataset cannot be easily inferred, thereby protecting user privacy. The introduction of noise acts as a privacy-preserving mechanism by obfuscating the exact contribution of each user's data, making it difficult for adversaries to reverse-engineer the original inputs from the noisy outputs.

McMahan et al. [39] pioneered the concept of user-level differential privacy (DP) to ensure robust privacy in federated learning. The primary goal of user-level DP is to make it impossible for the server or external adversaries to deduce whether any particular user participated in the federation. This is crucial in scenarios where the mere participation of a user could reveal sensitive information. The process includes several critical steps:

- **Step 1.** The server initializes the model parameters and selects a batch of clients to start the federated training process. This selection is often done randomly or based on specific criteria such as data quality or network conditions.
- **Step 2.** The selected clients request and download the model parameters from the server. They then train their local models based on their local datasets for several epochs. This local training phase allows the model to learn from the diverse and distributed data available across the different clients.
- **Step 3.** Once local training is completed in parallel, the clients clip their gradients and transmit them to the server. Clipping the gradients involves limiting their

norm to a pre-defined threshold, which helps in controlling the influence of any single data point on the overall model.

- **Step 4.** The server aggregates the clipped gradients using bounded-sensitivity estimators and updates the global model parameters by adding Gaussian noise. The addition of Gaussian noise is calibrated based on the privacy budget (ϵ, δ), ensuring that the aggregated gradients do not reveal individual data points.
- **Step 5.** Steps 2–4 are repeated until the model converges. The iterative nature of this process allows the model to gradually improve while maintaining strict privacy guarantees.

In practice, applying norm clipping can prevent potential gradient explosion [64] and ensures that each individual sample's influence on the aggregated results is bounded [1]. Norm clipping is a critical step in maintaining privacy and stability during the federated learning process as it limits the impact of any single user's data on the model. By controlling the gradient norms, federated learning systems can prevent outlier data points from disproportionately affecting the model, which is essential for both privacy and model robustness.

Based on this process, DP-PrivRec [36] implements noisy stochastic gradient descent by drawing noise from the $(0, \sigma^2)$ Gaussian distribution. A numerically stable moment accountant is employed to calculate the privacy budget ϵ, providing an accurate estimate of the privacy loss over multiple iterations of the algorithm. This method ensures that the cumulative privacy budget remains within acceptable bounds, thereby maintaining robust privacy guarantees throughout the training process. The moment accountant technique is particularly useful in tracking the privacy loss over time, allowing for more precise control over the trade-off between privacy and model performance.

Contrary to adding noise at the server-side aggregation, many HFedRec systems [18, 26, 43, 51, 53, 58, 71] choose to add local noise to the gradients or model weights before submission. This approach adheres to the principles of local DP (LDP), ensuring that each user's data is perturbed before it leaves their device. Despite the increased noise and the additional perturbation introduced at each client, LDP achieves fine-grained privatization of local data. This method ensures that user data is protected even before it is aggregated, adding an additional layer of privacy protection. Local noise addition is particularly advantageous in scenarios where clients are concerned about data leakage even during the transmission of gradients or model updates to the server.

Considering the varying magnitudes of gradients, some advanced techniques, such as FeSoG [32], and DFSR [48], have been developed to devise adaptive noisy strength based on $mean(\nabla_g)$. These techniques dynamically adjust the amount of noise added to the gradients based on their statistical properties, helping to balance the trade-off between privacy and utility. Furthermore, FedNCF [56] sets different thresholds to clip gradients, helping manage the trade-off between privacy and model accuracy. Adaptive noise strength helps in adjusting the amount of noise based on the magnitude of the gradients, preserving the utility of the model while ensuring robust privacy guarantees. This approach is particularly useful in federated learning environments where the distribution of data and the resulting gradients can vary significantly across different clients.

These methods collectively enhance the privacy and performance of federated learning systems. By integrating these techniques, federated learning can maintain high levels of model performance while effectively protecting user privacy. Continued research and innovation in this area are crucial for advancing privacy-preserving machine learning and ensuring that sensitive user data remains secure in distributed learning environments. As federated learning becomes more widely adopted, the development of sophisticated additive noise techniques will be essential for balancing the competing demands of privacy and model utility, ultimately leading to more secure and effective federated learning systems.

3.2.2 Randomized Response

Randomized response (RR) protects original data through the uncertainty when responding to sensitive topics [20]. In other words, for ID-sensitive rating behaviors, when user i interacted with item j, there is a probability to reverse the fact. This technique introduces deliberate randomness into users' responses, thereby providing plausible deniability and protecting individual privacy. By ensuring that individual responses are randomized, RR mitigates the risk of reidentification and ensures that sensitive information is not disclosed. The core principle of RR lies in its ability to obscure the truth in a controlled manner, thus making it difficult for adversaries to infer specific details about any individual user from the data.

Studies such as [4, 61] perturb the user-item interaction matrix by two-stage randomization including a permanent RR (PRR) and an instantaneous RR (IRR). PRR perturbs the binary matrix before model training and thus is termed "permanent," while IRR is applied to the binary matrix at each communication round. PRR ensures that the initial dataset used for training is already anonymized, thus maintaining privacy from the outset. On the other hand, IRR introduces additional noise during each round of communication between clients and the server, thereby enhancing privacy protection throughout the training process. The combination of PRR and IRR ensures that privacy is preserved both at the start and throughout the federated learning life cycle. By employing both PRR and IRR, these methods provide a layered defense against privacy breaches, making it increasingly difficult for adversaries to extract meaningful information.

Similarly, FedIS first obfuscates real interaction labels and then performs sparsification of model updates with the Bernoulli distribution. By obfuscating interaction labels, FedIS ensures that individual user actions cannot be directly inferred from the training data. The subsequent sparsification process further reduces the risk of privacy breaches by limiting the amount of information shared in each communication round. This dual approach of obfuscation and sparsification provides a robust mechanism for maintaining privacy while still enabling effective model updates. The combination of these techniques allows for the preservation of user privacy even in scenarios where the interaction data is sparse or highly sensitive.

Due to the perturbation enlargement phenomenon [34] in heterogeneous information networks (HINs), FedHGNN [53] utilizes the exponential mechanism and degree-preserving RR [69] to prevent a denser perturbed graph. The perturbation enlargement phenomenon refers to the issue where perturbations in HINs can lead to a significant increase in the density of the graph, which can degrade the quality of the model. To mitigate this, FedHGNN applies a degree-preserving RR, which ensures that the overall structure of the network is maintained while introducing sufficient randomness to protect privacy. The server can reconstruct a global HIN from local perturbed results while ensuring privacy security. This reconstruction process allows the global model to leverage the full breadth of the data without compromising individual user privacy. The application of degree-preserving RR is particularly important in scenarios where the structural integrity of the data is crucial for maintaining model performance.

Furthermore, the use of RR in federated learning settings provides a robust mechanism to balance privacy and data utility. By carefully calibrating the parameters of the RR mechanisms, it is possible to achieve high levels of privacy protection while still maintaining the accuracy and effectiveness of the federated learning model. The incorporation of RR into federated learning protocols represents a significant advancement in the field, offering a practical and scalable solution for privacy-preserving data analysis. By adjusting the parameters of the RR process, federated learning systems can fine-tune the balance between privacy and utility to meet specific needs and constraints. The flexibility of RR allows it to be tailored to different federated learning scenarios, making it a versatile tool for a wide range of applications.

The applied RR techniques in [4, 53, 61] also demonstrate various methods to balance the trade-offs between privacy and utility. These studies show how different configurations and applications of RR can be fine-tuned to meet specific privacy requirements without significantly sacrificing the performance of the federated learning models. This adaptability makes RR a versatile tool in the privacy-preserving arsenal of federated learning systems. By exploring different perturbation strategies and optimizing the application of RR, these methods ensure that federated learning can remain both secure and effective. The ability to balance these trade-offs is critical for the success of federated learning systems, especially in environments where privacy concerns are paramount.

Moreover, the flexibility of RR allows it to be integrated with other privacy-preserving techniques. For instance, RR can be combined with differential privacy (DP) or homomorphic encryption (HE) to further enhance privacy guarantees. This multilayered approach to privacy ensures that even if one mechanism is compromised, additional layers of protection remain in place. The synergy between RR and other techniques provides a comprehensive framework for safeguarding user data in federated learning. By layering different privacy-preserving methods, federated learning systems can achieve a more robust and resilient defense against potential privacy threats.

In summary, randomized response techniques enhance the privacy of federated learning systems by introducing uncertainty and randomness into the data. This approach ensures that individual user actions and preferences are obfuscated, thereby protecting user privacy while still allowing for effective and accurate model training. As the field continues to evolve, further refinements and optimizations of RR techniques will be essential to address emerging privacy challenges and enhance the robustness of federated learning systems. The continued development of RR methods will play a crucial role in ensuring that federated learning can be conducted securely and effectively in increasingly complex and privacy-sensitive environments. By integrating RR with other privacy-preserving techniques and optimizing its application, federated learning systems can achieve a robust balance between privacy and utility, ensuring the protection of user data while maintaining the effectiveness of the models.

3.2.3 Summary and Challenges

Compared with HE, DP has lower computational complexity and can prevent privacy leakage while keeping higher efficiency, such as nonzero transmission in FedIS's sparsity perturbation. This makes DP particularly attractive for federated learning scenarios where computational resources and bandwidth are often limited. However, the introduction of noise into the training process can negatively impact model accuracy. The degree of privacy preservation is inversely proportional to model accuracy, presenting an intrinsic challenge in how to trade off security and accuracy [18, 51, 58]. Ensuring rigorous theorization for the guarantee of (ϵ, δ)-differential privacy in both training and online serving stages is critical to addressing this challenge [3]. A careful balance must be struck to ensure that the noise added to protect privacy does not excessively degrade the utility of the resulting models.

Furthermore, due to the intuitive composition theorems and compatibility with Gaussian noise, Rényi differential privacy (RDP) [40] is increasingly receiving much research attention [10, 17, 35]. DGREC [12] pioneers the investigation of RDP-based secure gradient-sharing with tighter bounds on the privacy budget. RDP offers a refined approach to traditional differential privacy by leveraging Rényi divergence, which generalizes the Kullback-Leibler divergence to provide more nuanced privacy guarantees.

At the heart of RDP is the Rényi divergence, a measure of the difference between two probability distributions that generalizes the Kullback-Leibler divergence. Given two probability distributions P and Q, and a parameter $\alpha > 1$, the Rényi divergence of order α, denoted as $D_\alpha(P \parallel Q)$, is defined as follows:

$$D_\alpha(P \parallel Q) = \frac{1}{\alpha - 1} \log \left(\sum_x P(x)^\alpha Q(x)^{1-\alpha} \right) \tag{3.4}$$

Rényi differential privacy uses this divergence to quantify the privacy loss of a mechanism. Specifically, a mechanism M is said to satisfy (α, ϵ)-RDP if, for any two neighboring datasets D and D' that differ in a single element, the Rényi divergence between the output distributions of M on D and D' is bounded by ϵ:

$$D_\alpha(M(D) \parallel M(D')) \leq \epsilon \tag{3.5}$$

This flexibility enables detailed privacy analysis and better utility, making RDP suitable for federated learning. By using Rényi divergence, RDP provides a more refined and tighter analysis of privacy loss, which can result in more accurate accounting of the privacy budget over multiple computations. This is particularly important in federated learning, where multiple rounds of communication and aggregation occur, and traditional DP methods may overestimate the privacy loss, leading to overly conservative and inefficient noise addition.

Moreover, RDP's compatibility with Gaussian noise is another significant advantage. Gaussian noise is often preferred in practice due to its desirable mathematical properties and ease of implementation. The ability of RDP to naturally integrate with Gaussian noise means that it can be seamlessly applied to many existing DP mechanisms, enhancing their effectiveness and efficiency. The integration of Gaussian noise with RDP allows for more accurate control over the privacy-utility trade-off, ensuring that the noise addition process does not excessively compromise model performance while maintaining strong privacy guarantees.

In summary, the adoption of RDP in federated learning represents a significant advancement in the quest for achieving robust privacy preservation while maintaining high model utility. As research in this area continues to evolve, it is expected that further refinements and optimizations of RDP will emerge, providing even more powerful tools for balancing privacy and accuracy in federated learning systems. More refined and quantitatively tractable analyses on privacy loss remain to be studied. The continuous development of RDP and its integration into federated learning frameworks will be essential for advancing the field of privacy-preserving machine learning, ensuring that sensitive user data is protected while enabling effective and efficient model training.

Additionally, practical implementation challenges must be addressed to fully realize the potential of RDP in federated learning. These include optimizing the computational efficiency of RDP algorithms, developing scalable solutions for large-scale federated systems, and ensuring compatibility with various data types and model architectures. Researchers and practitioners must work together to create robust, scalable, and efficient RDP implementations that can be seamlessly integrated into existing federated learning frameworks. For example, optimizing the implementation of RDP in edge computing environments, where computational resources are constrained, is critical for ensuring that privacy-preserving techniques do not hinder real-time processing capabilities.

Future work should also focus on developing adaptive privacy-preserving mechanisms that can dynamically adjust their parameters based on the evolving needs of the federated learning environment. Such adaptive mechanisms can provide more flexible and responsive privacy protections, ensuring that the system can maintain high levels of utility even as the data and model requirements change over time. This could involve the development of privacy-preserving techniques that adjust the level of noise added based on the sensitivity of the data or the importance of the model's task, allowing for a more nuanced and effective trade-off between privacy and accuracy. By continuously refining and enhancing these privacy-preserving techniques, the federated learning community can build more secure, effective, and user-friendly systems that meet the growing demand for privacy-conscious data analysis and machine learning.

In addition to these technical advancements, there is also a need for comprehensive evaluations of RDP's effectiveness in real-world federated learning scenarios. Such evaluations should consider not only the theoretical privacy guarantees but also the practical implications of implementing RDP in diverse environments, including its impact on model accuracy, system performance, and user experience. Developing standardized benchmarks and metrics for assessing the performance of RDP in federated learning will be crucial for guiding future research and ensuring that these techniques are both practical and effective in safeguarding user privacy.

In conclusion, while RDP offers a powerful solution for enhancing privacy in federated learning, its practical adoption requires addressing several key challenges, including computational efficiency, scalability, and adaptability. By focusing on these areas, the federated learning community can ensure that RDP remains a viable and effective tool for securing sensitive data in distributed machine learning environments. The ongoing research and development in this field will be pivotal in shaping the future of secure, efficient, and scalable federated learning systems, paving the way for broader adoption of these privacy-preserving technologies in real-world applications. As federated learning continues to gain traction across various industries, the importance of robust and scalable privacy-preserving techniques like RDP will only increase, underscoring the need for continued innovation and collaboration in this area.

3.3 Secure Multi-party Computation

Secure multi-party computation (SMC) is a general cryptographic scheme that enables a group of participants to cooperatively compute an objective function without exposing their private data in a distributed environment [33]. Participants solely acquire their own corresponding computational result (if needed) and have no awareness of other participants' private data. This method ensures that sensitive information remains confidential, even while being used for collaborative computation. SMC has become increasingly relevant in the context of federated learning, where the need to protect user data while enabling collaborative model training is paramount.

SMC first entered the public view through Yao's groundbreaking work on Garbled Circuits [59]. Since then, it has incorporated multiple cryptographic tools after decades of development, including the Secret Sharing (SS) protocol and the Garbled Circuit protocol. These protocols provide different mechanisms for achieving secure computation while maintaining data privacy. The evolution of SMC has led to a diverse set of techniques that can be applied depending on the specific needs and constraints of the federated learning task.

Compared to HE-based and DP-based HFedRec, SMC-based HFedRec offers a broader range of privacy-preserving protocols. These include the Secret Sharing protocol, Garbled Circuit protocol, and Oblivious Transfer (OT) protocol. Each of these protocols has its own unique advantages and can be chosen based on the specific requirements of the federated learning task. SMC's versatility in offering multiple secure computation options makes it a powerful tool for privacy-preserving federated learning.

3.3.1 Secret Sharing

Secret Sharing means that a dealer fragments private data called *secret* into pieces of *shares* distributed among participants [45]; Fig. 3.4 shows the application of Secret Sharing. The original secret can be reconstructed by combining sufficient shares. In the federated setting, the split secret generally refers to publicly transmitted model parameters. To prevent secrets from divulging privacy to an honest-but-curious server or attackers, the *secure aggregation* problem [8, 49] is termed to build a trustworthy aggregation schema. The server is blind to what clients submit and only acquires aggregated results. This ensures that individual data points remain confidential while still allowing the server to compute the necessary global updates.

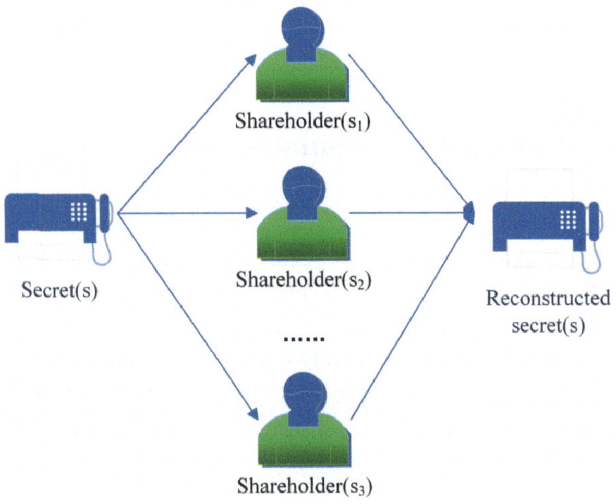

Fig. 3.4 Secret sharing

An intuitive practice is secret splitting [47], where each client divides a gradient ∇_g (or model weights) into n parts, and all sub-gradients are required to recover $\nabla_g = \sum_{i=1}^{n} \nabla_{g_i}$. Local item gradients are first divided into several random numbers, one of which is retained locally, and then the rest are distributed to others [25, 57]. If there are clients A and B to upload their gradients X and Y, respectively, the sharing steps are as follows:

- **Step 1.** Clients A and B randomly divide their gradient into $\{x_1, x_2\}$ and $\{y_1, y_2\}$, respectively. This initial division ensures that the original gradient values are never fully exposed at any single point in the process.
- **Step 2.** Clients A and B distribute their sub-gradients to each other. Client A holds $\{x_1, y_1\}$, and client B holds $\{x_2, y_2\}$. By sharing these sub-gradients, both clients ensure that neither has a complete view of the other's data.
- **Step 3.** Both sum locally stored sub-gradients, i.e., $z_1 = x_1 + y_1$ in Client A, and $z_2 = x_2 + y_2$ in Client B. z_1 and z_2 are uploaded to the central server. These locally computed sums represent partial contributions to the global model update.
- **Step 4.** The server gets the aggregated gradient $Z = z_1 + z_2$. The final gradient used for the global model update is thus an aggregation of the contributions from both clients, with no individual data point being fully exposed.

Lin et al. [31] further propose a generic cross-user HFedRec framework that examines more MF variants and neural networks under the above SS process. As a single update gradient is divided into multiple shares, the server may misestimate the real gradient number, leading to suboptimal averaging results. FMFSS [50] simultaneously leverages the splitting operation to securely derive user-agnostic item interaction frequencies at the server side. This approach enhances privacy by ensuring that sensitive user information is not directly linked to the specific data points used in model training.

However, forcing all shares to reconstruct the original information is impractical when clients join and drop out of the federation dynamically, i.e., some shares may be lost before aggregation. More robust SS protocols are necessary to deal with the unpredictable client activities. One such protocol is Shamir's Threshold Secret Sharing scheme [45]. This scheme controls the number of shares required to reconstruct the secret, providing flexibility and robustness in dynamic environments. Shamir's scheme is particularly well-suited to federated learning scenarios where participant availability can vary over time.

Shamir's Threshold Secret Sharing scheme [45] is based on polynomial interpolation. The core idea is to encode the secret as the constant term of a polynomial of degree $t - 1$, where t is the threshold number of shares required to reconstruct the secret. Each share is a point on this polynomial, and any t shares can be used to reconstruct the polynomial and thus the secret. Specifically, the dealer generates a random polynomial $P(x) = a_0 + a_1x + a_2x^2 + \ldots + a_{t-1}x^{t-1}$, where a_0 is the secret. The dealer then evaluates the polynomial at n different points to create the shares and distributes them to the participants. The secret can be reconstructed by interpolating the polynomial using any t shares. This approach allows for secure data reconstruction even when some participants drop out, making it ideal for federated learning environments.

Bonawitz et al. [7] propose a well-designed secure aggregation schema based on Shamir's SS. This schema meets multiple requirements, including secure summation, constant communication rounds, low communication overhead, and failure robustness. Many HFedRec systems [21, 30, 37, 65] are motivated by their work to implement parameter aggregation in a secure manner. For example, in order to hide user-item interaction behaviors, efficient FedRec [37] also utilizes Bonawitz's solution to unite news sets from a user group before the gradient aggregation. By applying these advanced Secret Sharing techniques, federated learning systems can ensure that sensitive user data is protected while still enabling effective collaborative learning.

In conclusion, Secret Sharing protocols, particularly Shamir's Threshold Secret Sharing, provide a robust mechanism for secure aggregation in federated learning environments. By enabling the reconstruction of secrets only when a sufficient number of shares are combined, these protocols ensure the privacy and integrity of the data even in dynamic and potentially adversarial settings. As federated learning continues to evolve, the integration of these advanced cryptographic techniques will be essential for maintaining data privacy and enhancing the security of collaborative learning frameworks. The continued development and application of Secret Sharing protocols in federated learning will play a crucial role in enabling secure, privacy-preserving data analysis in increasingly complex and distributed environments.

3.3.2 Garbled Circuit

Garbled Circuit initially involves secure two-party computation in Yao's protocol [59]. The word "Circuit" means that parties jointly compute a public function expressed as a logic circuit without exposing individual input data. The word "Garbled" refers to the garbled truth table in which each encrypted entry is randomly ordered and embodies inputs and the corresponding output of a gate [6]. In a Garbled Circuit, the function to be computed is represented as a Boolean circuit, and each gate in the circuit is "Garbled" in such a way that the inputs and outputs remain encrypted, ensuring that the actual data values are not revealed during computation. This encryption ensures that sensitive data remains protected throughout the entire computation process, making Garbled Circuit (GC) a foundational technique for privacy-preserving secure multi-party computation (SMC).

The communication cost primarily depends on the size of the garbled truth table, and the computation complexity lies in the encryption and decryption operations on the table's entries. As a result, the efficiency of GC protocols is closely tied to the optimization of these tables and the cryptographic operations involved. Subsequent GC protocols have focused on decreasing communication and computation costs [27, 62] and generalizing the scale to multiple parties [5, 19]. These advancements have significantly broadened the applicability of GC in various secure computation scenarios.

For example, the FreeXOR technique [27] allows XOR gates to be evaluated without any cryptographic operations, significantly reducing the computational overhead. This technique is particularly effective in circuits where XOR gates are prevalent, as it eliminates the need for expensive encryption and decryption steps for these operations. The FreeXOR method thus enhances the efficiency of GC protocols, making them more practical for real-world applications.

The Half-Gate technique [62] further optimizes the evaluation of AND gates, which are more complex than XOR gates, by reducing the number of necessary cryptographic operations. AND gates typically require more resources to garble and evaluate, so the Half-Gate technique's ability to streamline this process is crucial for improving the overall performance of GC protocols. By reducing the cryptographic burden associated with AND gates, the Half-Gate method allows for more scalable and efficient secure computations.

In multi-party settings, the BMR protocol [5] extends the Garbled Circuit approach to securely compute functions over inputs from multiple parties. This protocol enables secure multi-party computation by distributing the garbling and evaluation tasks among several participants, thereby increasing the scalability and efficiency of the protocol. The BMR protocol's ability to handle inputs from multiple parties makes it well-suited for complex federated learning tasks where data is distributed across different entities.

The Half-Gate technique has also been adapted for multi-party computation, further improving the performance of GC protocols in such environments [19]. By applying the optimizations of the Half-Gate technique to multi-party settings, these protocols can efficiently handle the secure evaluation of Boolean circuits involving multiple participants. This extension of GC protocols to multi-party scenarios is particularly important for federated learning, where data privacy must be maintained across multiple, potentially untrusted, entities.

From our observation, only FederatedCF [24] explores GC and involves two organizations of which one undertakes the garbler and the other undertakes the evaluator. It forms the classic two-party GC schema and performs Yao sharing on user behaviors. In this setting, one party (the garbler) constructs the Garbled Circuit and encrypts the inputs, while the other party (the evaluator) evaluates the Garbled Circuit without learning the actual inputs. This approach ensures that the computation is carried out securely, with neither party gaining access to the other's private data. The secure two-party computation provided by GC in FederatedCF demonstrates the protocol's effectiveness in maintaining data privacy while allowing for collaborative computations.

FederatedCF demonstrates the practical application of GC in federated learning by securely computing collaborative filtering recommendations. By leveraging the GC protocol, FederatedCF ensures that user preferences and interactions are kept confidential while still allowing the collaborative filtering model to be trained effectively. This use of GC highlights the potential for secure two-party computation protocols to enhance privacy in federated learning systems. The successful implementation of GC in FederatedCF showcases its viability for privacy-preserving machine learning applications, particularly in scenarios where data confidentiality is paramount.

Moreover, as federated learning continues to evolve, the integration of advanced GC techniques can further enhance the privacy and security of collaborative computations. Future research may focus on optimizing GC protocols for federated learning by reducing the communication and computation overhead and extending the protocols to support more complex machine learning algorithms. By addressing these challenges, GC can play a crucial role in enabling secure and privacy-preserving federated learning at scale. Continued innovation in this area will be essential for ensuring that federated learning can effectively balance the need for data privacy with the demands of large-scale, distributed computation.

In conclusion, Garbled Circuit protocols offer a powerful framework for secure computation in federated learning. By encrypting the inputs and outputs of logical gates, GC ensures that sensitive data remains confidential throughout the computation process. The continued development and optimization of GC techniques will be essential for advancing the field of privacy-preserving federated learning and enabling secure collaborative computations across diverse and distributed datasets. As federated learning becomes more widespread, the role of GC in ensuring the security and privacy of these computations will become increasingly important, paving the way for more secure and effective machine learning models.

3.3.3 Summary and Challenges

SMC aims at multi-party privacy security and is naturally adaptive to HFedRec. The secure aggregation schema can generate equivalent results as vanilla aggregation [65]. This ensures that the privacy-preserving mechanisms do not compromise the quality or accuracy of the aggregated results. However, one growing concern is how to reduce the multi-party communication cost (e.g., data volume, time, rounds) as the client size increases [30, 37, 65]. As the number of participants in federated learning increases, the communication overhead can become a significant bottleneck. Efficient protocols that minimize the data exchanged and the number of communication rounds are essential to ensure scalability.

Moreover, recent HFedRec research has only explored the tip of the iceberg in the SMC field. While Secret Sharing (SS) and Garbled Circuit (GC) protocols are commonly used, there are many other SMC protocols that could enhance privacy and security in federated learning. For instance, protocols like SS for confronting dishonest dealers provide mechanisms to ensure that even if some participants act dishonestly, the overall computation remains secure. These protocols can detect and mitigate the impact of dishonest participants, ensuring the integrity of the computation.

Garbled Circuit (GC) protocols for defending against malicious adversaries present another promising area of exploration. The Cut-and-Choose technique, for example, enhances the security of GC by randomly sampling and verifying parts of the circuit, thereby detecting and preventing malicious behavior. This technique can be adapted to federated learning to protect against sophisticated attacks where adversaries might attempt to manipulate the computation by introducing malicious inputs.

In addition to addressing these specific challenges, further research is needed to integrate and optimize these advanced SMC protocols within the federated learning framework. This includes developing efficient algorithms that can leverage the strengths of various SMC techniques while minimizing their weaknesses. For example, hybrid approaches that combine the simplicity of SS with the robustness of GC could provide a balanced solution for secure and efficient federated learning. These hybrid methods would be particularly valuable in scenarios where the simplicity of SS is needed to reduce computational overhead, while the robustness of GC is required to defend against more sophisticated threats.

Another critical area of research is the development of adaptive protocols that can dynamically adjust their security and efficiency parameters based on the current state of the network and the behavior of participants. Such protocols could provide enhanced security during periods of high activity or when suspicious behavior is detected while relaxing the security requirements to improve efficiency during normal operation. This adaptability is crucial for maintaining a balance between security and performance, especially in dynamic and large-scale federated learning environments where conditions can change rapidly.

Furthermore, the scalability of SMC protocols is a significant challenge that needs to be addressed to ensure their applicability in large-scale federated learning scenarios. As the number of participants grows, the communication and computation costs of traditional SMC protocols can become prohibitive. Therefore, developing scalable SMC protocols that can efficiently handle thousands or even millions of participants is essential. This might involve exploring novel communication strategies, such as hierarchical aggregation or federated approaches that reduce the complexity of multi-party interactions.

Additionally, there is a growing need to investigate the robustness of SMC protocols against new and emerging threats in federated learning. As adversaries become more sophisticated, they may develop novel attack vectors that exploit weaknesses in existing SMC protocols. To counter these threats, researchers must continuously evolve SMC techniques, incorporating cutting-edge cryptographic advances and developing innovative defenses that can anticipate and neutralize these attacks before they can cause significant harm.

The integration of SMC with other privacy-preserving technologies, such as differential privacy (DP) and homomorphic encryption (HE), also presents an intriguing area for future exploration. Combining these technologies could potentially offer a more comprehensive security framework, addressing the limitations of each approach while enhancing overall system robustness. For instance, using HE to secure the data before applying SMC protocols could provide an additional layer of protection, while DP could be used to ensure that the outputs of federated learning models do not reveal sensitive information about individual participants.

In conclusion, while SMC provides a robust foundation for privacy-preserving federated learning, there are significant challenges and opportunities for further research. By addressing the issues of communication cost, scalability, and robustness against dishonest and malicious participants, the full potential of SMC in federated learning can be realized. The continued evolution and integration of advanced SMC protocols will be crucial for ensuring that federated learning systems remain secure, efficient, and scalable in increasingly complex and adversarial environments. The future of federated learning will likely depend on the successful adaptation and enhancement of SMC techniques, ensuring that these systems can meet the growing demands for privacy, security, and performance in a wide range of real-world applications. As the field progresses, collaborative efforts between cryptographers, data scientists, and system engineers will be key to developing the next generation of federated learning systems that are both secure and scalable.

3.4 Malicious Attacks and Byzantine Resilience

Despite the federated security augmented by privacy-preserving techniques, user privacy and recommendation performance are still susceptible to malicious attacks [28]. Parties in the aforementioned HFedRec systems are generally assumed to passively obey the promissory federated process. However, even if a small percentage of active adversaries tamper with the federation, stealthy and harmful attacks can compromise the model validity and expose private user data. We have organized the attack and defense studies into Table 3.1. In addition to specific attack assumptions, there are several common assumptions for all attackers:

(1) They can compromise a small proportion of clients if attacks are from malicious clients. This means that only a few clients need to be under the control of the adversary to launch effective attacks.
(2) They have prior knowledge of a candidate item pool, especially when the item embedding matrix is transmitted between the server and clients. This allows attackers to target specific items or groups of items.
(3) They have no access to benign clients' local data, such as interacted items and user embeddings. Whether malicious clients can manipulate their own private data depends on the specifics of the attack.

In the following subsections, we trace recent progress in understanding and defending against poisoning and inference attacks.

3.4.1 *Poisoning Attack*

Poisoning attacks in federated learning can have devastating effects on both the integrity of the model and the privacy of the data. These attacks are typically

Table 3.1 Representative works of attack and defense

Publications	Type	Model	Constraints	Special attack assumes
FedRecAttack [38] **Aim:** ItemPromotion	Targeted model poison on uploaded gradients	BPRMF with LDP	Limit the maximum l2-norm and number of poisoned item gradients	**Access** the model structure (e.g., hidden layers' weights and item embeddings) except all user embeddings **Access** some hyperparameters (e.g., learning rate) **Have** a small fraction of public user-item interactions **Alter** malicious clients' gradients
PipAttack [66] **Aim:** ItemPromotion	Targeted model poison on uploaded gradients	NCF	Limit the distance between poisoned gradients and the original ones	**Access** the model structure except benign user embeddings **Have** clues on the popularity of items **Alter** malicious clients' local models and their gradients
A-ra and A-hum [44] **Aim:** ItemPromotion	Targeted model poison on uploaded gradients	NCF	–	**Access** the model structure except all user embeddings **Alter** malicious clients' gradients
PSMU [52] **Defenses:** HiCS **Aim:** ItemPromotion	Targeted model poison on uploaded gradients	NCF and LightGCN	Only upload the target items' poisoned gradients	**Access** the model structure except benign user embeddings malicious clients are synthetic. Read-write privileges, such as local data and user embeddings, are naturally owned)

(continued)

Table 3.1 (continued)

Publications	Type	Model	Constraints	Special attack assumes
FedAttack [29] **Aim:** ValidityFailure	Untargeted data poison on pos./neg. samples	NCF and BERT4Rec	–	**Access** local user and item embeddings **Have** normal user profiles to infer user embeddings **Alter** malicious clients' local model inputs and labels
ClusterAttack [13] **Defense:** UNION **Aim:** ValidityFailure	Untargeted model poison on uploaded gradients	BPRMF and SASRec	Clip poisoned gradients with a norm bound estimated from normal gradients	**Access** the model structure except all user embeddings **Have** normal interactions to clip poisoned gradients **Alter** malicious clients' gradients
IMIA [23] **Defense:** RegTerm **Aim:** PrivateData	Membership inference on user-item interactions	NCF and LightGCN with LDP	–	**Access** the target clients' uploaded parameters in accord with pre-agreed federated protocol **Access** some hyperparameters (e.g., negative sampling rate and learning rate)
EIFedMF [22] **Aim:** PrivateData	Defense against attribute inference attack	PMF with LDP, HE, and OT	–	–

Aim: ItemPromotion: the attack aims to popularize target items as many users as possible. ValidityFailure: the attack aims to degrade the model performance. PrivateData: the attack aims to infer private user data (interacted items in IMIA and user profiles in EIFedMF)
Defense: the proposed defense method to address the corresponding attack. **Type:** Attack and defense types
Model (victim example)**:** BPRMF, Bayesian personalized ranking-based MF; NCF, neural collaborative filtering; LightGCN, Light GCN; SASRec Self-Attentive Sequential Recommendation; BERT4Rec, BERT-based recommendation; PMF, probability MF; OT, Oblivious Transfer [15]
Constraints: to prevent the attack from being detected (i.e., stealthy) and affecting the model performance; **Assumes**: prior knowledge of each attack model

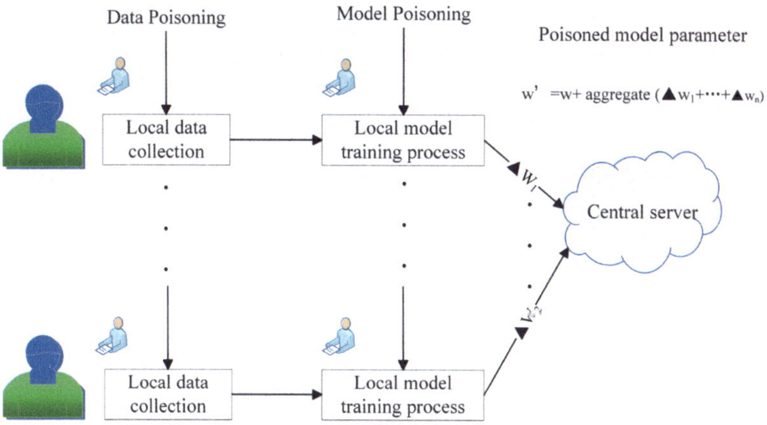

Fig. 3.5 Poisoning attacks

categorized into two main aims: item promotion and validity failure. Each of these aims targets different aspects of the federated learning process, either by manipulating model outcomes to favor certain items or by degrading the overall model performance.

The first aim, item promotion, seeks to boost target items' exposure rates without negatively impacting the overall model performance. This type of attack, known as targeted poison, shifts users' interests toward items specified by adversaries. Due to the inherent popularity bias in recommended items, adversaries can exploit this bias to increase the visibility of specific items. For instance, PipAttack [66] conceals target items through a technique known as popularity obfuscation; Fig. 3.5 shows the application of poisoning attacks. This approach involves training an additional classification model to label each item with a popularity level, aligning target items with those that are already top-popular, thereby increasing their exposure in recommendations.

Another example of an effective item promotion attack is FedRecAttack [38], which utilizes publicly available user-item interactions to approximate user embeddings. The experiment demonstrates that even with only 3% malicious clients and 1% public interactions, high effectiveness can be maintained. This finding underscores the vulnerability of federated learning systems to even a small number of adversarial inputs, which can significantly influence the model's behavior. Subsequent studies have attempted to achieve similar effects with less prior knowledge. For example, Rong et al. [44] propose a method that randomly approximates user embeddings from a Gaussian distribution. However, the performance of this attack is less stable due to the randomness involved. To address this, PSMU [52] constructs synthetic clients with fake interactions. Target items that are highly ranked in synthetic clients exhibit a popularity bias among benign clients as well. This method effectively amplifies the predicted scores of target items relative to their substitutes and top-K recommendations. In response, a corresponding defense method is proposed that involves hierarchical gradient clipping and sparsified updating at the server, thereby reducing the impact of malicious gradients on the global model.

The second aim, validity failure, is focused on undermining the overall performance of the federated model. This type of attack, known as untargeted poison, affects users' normal experience of recommendation services by degrading the accuracy and reliability of the model. ClusterAttack [13] is a prominent example of this approach. The attack begins by applying k-means clustering to item embeddings received from the server. A within-cluster variance loss is then computed to minimize the discrepancy between item embeddings and their centroids, which perturbs the actual item embedding distribution and degrades the model's validity. The cluster number k is adaptively decided based on the loss variation, allowing the attack to dynamically adjust to the model's structure. As a defense against this attack, a contrastive task is adopted to sustain the uniform distribution of item embeddings, thereby preserving the model's integrity and preventing the degradation caused by ClusterAttack.

All of the above poisoning attacks are classified as model poisoning attacks, where adversaries manipulate the gradients uploaded to the server to influence the model. These attacks are particularly insidious because they do not require direct manipulation of the training data itself; instead, they exploit the federated learning process by altering the updates that are aggregated to form the global model. However, another type of poisoning attack, known as data poisoning, involves direct manipulation of the training data. FedAttack [29] is a notable example of this approach, where adversaries subvert HFedRec by reversing the labels of positive and negative items. This type of attack introduces incorrect data into the training process, leading the model to learn erroneous patterns and degrading its overall performance.

Recent research efforts have expanded the scope of poisoning attacks to include more sophisticated techniques. For example, Zhang et al. have introduced a hybrid approach that combines both model poisoning and data poisoning. In this method, attackers manipulate gradients while simultaneously injecting adversarial examples into the training data. This dual approach significantly enhances the attack's efficacy, making it more challenging to detect and mitigate. The combination of model and data manipulation allows adversaries to exert a greater influence on the model's behavior, leading to more severe disruptions in the federated learning process.

Additionally, a novel framework called Adversarial Federated Learning (AFL) has been proposed to systematically evaluate the robustness of federated learning models against poisoning attacks. AFL provides a comprehensive benchmark by incorporating various attack vectors, including gradient manipulation, data injection, and backdoor attacks. This framework aids in understanding the vulnerabilities of federated learning systems and developing more resilient defense mechanisms. By providing a structured approach to evaluating and defending against poisoning attacks, AFL helps researchers and practitioners better safeguard federated learning models from adversarial threats.

In summary, poisoning attacks in federated learning pose significant threats to both the performance and integrity of recommendation systems. While targeted poison attacks aim to promote specific items, untargeted poison attacks seek to disrupt the overall model performance. Effective defenses against these attacks require robust strategies such as gradient clipping, sparsified updating, and contrastive tasks to mitigate the influence of malicious inputs. The continuous evolution of attack strategies necessitates ongoing research and development of advanced defense mechanisms to ensure the security and reliability of federated learning systems. As federated learning continues to be adopted in various applications, understanding and addressing these threats will be crucial for maintaining secure and reliable systems. The development of comprehensive evaluation frameworks like AFL will be instrumental in advancing the field and ensuring that federated learning models can withstand the growing sophistication of poisoning attacks.

3.4.2 Inference Attack

Inference attacks aim to extrapolate private information from training samples [46]. From our observation, recent inference attackers often involve a semi-honest server curious about private user data, whereas poisoning attackers are typically malicious clients. This distinction is important because it highlights the different motivations and methods used by adversaries in these contexts. While poisoning attackers seek to degrade the model or skew its outputs, inference attackers are more focused on extracting sensitive information from the data used in training.

IMIA [23] is a notable example of a membership inference attack. It first constructs a shadow dataset where labels of uploaded item embeddings are proportionally assigned by the negative sampling rate. The shadow model, trained on this dataset, mimics the behavior of the target model. Then, if an item embedding from the trained shadow model is similar to the uploaded one in Euclidean space, the membership inference on user behaviors is achieved. In this scenario, the attacker can determine whether a specific user-item interaction was part of the training data, thus compromising user privacy. The ability to infer whether a particular data point was used in training poses a significant privacy risk, as it can reveal sensitive information about individual users; Fig. 3.6 shows the application of inference attacks.

To defend against such attacks, a local defender can utilize a regularization term to limit the update of received public parameters, thereby reducing the attack's effectiveness. Regularization helps to mask the specific contributions of individual data points by introducing a degree of uncertainty into the model updates. This makes it more difficult for attackers to accurately determine membership, thus protecting user privacy.

To defend user profiles (e.g., age, gender) against attribute inference, EIFedMF [22] combines multiple privacy-preserving techniques. Full homomorphism of CKKS [14] is exploited to cluster user-item interactions securely. CKKS allows for encrypted computations on real numbers, enabling the clustering process to occur without exposing raw data. Following the clustering, local differential privacy

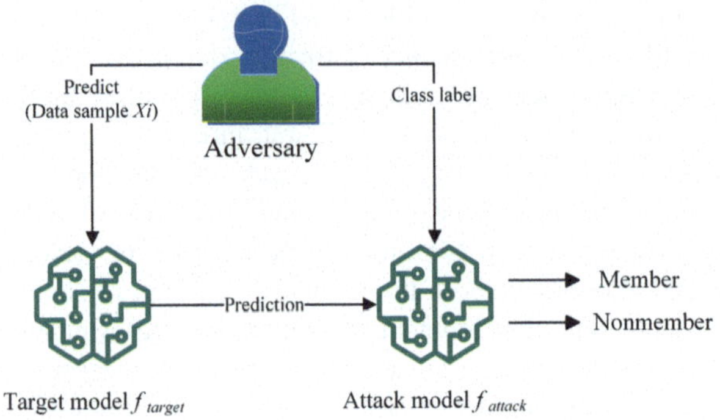

Fig. 3.6 Inference attacks

(LDP) bounds the probability of whether a user belongs to a particular cluster, adding an additional layer of privacy by introducing noise to the clustering process. The combination of CKKS and LDP provides a powerful defense against attribute inference attacks by ensuring that sensitive user attributes are protected at multiple stages of the learning process.

Finally, Oblivious Transfer (OT) [15] obfuscates a user's item ratings by transferring them from their own cluster to another. OT ensures that the server can transfer ratings without knowing which specific ratings are being transferred, thus preserving the privacy of individual users. This multifaceted approach combining CKKS, LDP, and OT provides robust protection against attribute inference attacks, ensuring that sensitive user attributes remain confidential even during the federated learning process. By integrating these techniques, EIFedMF offers a comprehensive defense strategy that addresses the various ways in which inference attacks can compromise user privacy.

Inference attacks pose a significant threat to the privacy of users in federated learning systems. These attacks can be broadly categorized into membership inference and attribute inference, each targeting different aspects of user data.

Membership inference attacks aim to determine whether a particular data point was included in the training dataset. In the context of federated learning, attackers exploit the updates sent to and from the central server to infer the presence of specific data points. IMIA [23] exemplifies this by constructing a shadow dataset and using similarity measures in Euclidean space to identify if certain user-item interactions were part of the training set. This can lead to severe privacy breaches, as attackers gain insights into individual user behaviors and preferences. Such attacks can be particularly damaging in recommendation systems, where user interaction data is often sensitive and personal.

To mitigate membership inference attacks, several defensive strategies can be employed. Regularization techniques, such as adding noise to model updates or using differential privacy mechanisms, can obscure the relationship between the updates and the underlying data. These methods reduce the ability of attackers to accurately determine membership by making the updates less reflective of individual data points. Additionally, the use of dropout or other stochastic techniques during training can further complicate the inference process for attackers, making it harder to pinpoint specific data points.

Attribute inference attacks, on the other hand, focus on deducing sensitive attributes of users (e.g., age, gender) from the shared model updates. EIFedMF [22] presents an innovative defense by integrating multiple privacy-preserving techniques. CKKS homomorphic encryption is utilized to perform secure clustering of user-item interactions, ensuring that the clustering process does not expose raw data. This allows for the creation of encrypted clusters that can be used for further analysis without compromising privacy. The use of homomorphic encryption in this context is particularly valuable because it allows computations to be performed on encrypted data, ensuring that sensitive information remains protected even during processing.

Local differential privacy (LDP) adds an additional layer of security by introducing randomness into the data before it is shared, thereby reducing the likelihood of accurate attribute inference. LDP ensures that the probability of any specific data point being part of a cluster is obfuscated, making it difficult for attackers to infer sensitive attributes with high confidence. The noise introduced by LDP makes it harder for attackers to draw accurate conclusions about individual users, thereby enhancing the overall privacy of the system.

Oblivious Transfer (OT) further enhances privacy by enabling the transfer of item ratings between clusters without revealing which ratings are being transferred. This ensures that even if the server is compromised, the specific ratings and their corresponding users remain confidential. By combining these techniques, EIFedMF provides a robust defense against attribute inference attacks, protecting user privacy throughout the federated learning process. The use of OT in this context helps to ensure that sensitive data remains protected even when it is being transferred between different parts of the system, providing an additional layer of security.

In summary, inference attacks represent a significant threat to the privacy of users in federated learning systems. These attacks exploit the shared model updates to infer sensitive information about user behaviors and attributes. Effective defenses require a combination of advanced cryptographic techniques and privacy-preserving mechanisms. Methods such as regularization, full homomorphic encryption, local differential privacy, and Oblivious Transfer are critical for protecting user privacy and maintaining the integrity of federated learning systems. By implementing these defenses, it is possible to mitigate the risks posed by inference attacks and ensure that federated learning models can be used safely and securely.

As the field of federated learning continues to evolve, ongoing research and innovation will be essential to develop even more robust defenses against these sophisticated attacks. Future advancements may include improved encryption schemes, more effective noise addition techniques, and enhanced protocols for secure data transfer. By staying ahead of potential threats, the federated learning community can ensure that user privacy is preserved while enabling the collaborative benefits of this powerful machine learning paradigm. The continuous development of new defense mechanisms will be crucial for keeping pace with the evolving nature of inference attacks and ensuring the long-term security of federated learning systems.

3.4.3 Summary and Challenges

Recent attack methods have achieved remarkable effectiveness in subverting the reliability and confidentiality of HFedRec. For instance, PSMU [52] can reach a 1.0 Exposure Rate@5 with only two or three malicious clients on MovieLens-1M, demonstrating the significant impact a small number of adversarial participants can have on the recommendation system. Similarly, IMIA [23] empirically reveals that local differential privacy (LDP) can preserve privacy; however, the additive noise introduced by LDP often significantly compromises recommendation performance. This indicates that there is an imbalance between existing privacy protection efforts and attack scenarios in HFedRec.

Despite the advances in privacy-preserving techniques, the current methods still struggle to provide a robust defense against sophisticated attacks. Attackers have been able to exploit vulnerabilities in federated learning systems to either promote specific items (targeted poison) or degrade overall system performance (untargeted poison), as well as to infer private user data (inference attacks). These challenges underscore the need for more comprehensive and resilient privacy protection mechanisms that can effectively counteract these attacks without sacrificing model accuracy.

One of the significant challenges in developing robust defenses is the dynamic and evolving nature of attack strategies. For example, hybrid attacks that combine elements of both model poisoning and data poisoning can be particularly difficult to detect and mitigate. Additionally, adversaries can employ adaptive strategies that change over time to bypass static defense mechanisms. This necessitates the development of adaptive privacy-preserving techniques that can dynamically adjust their parameters based on the observed behavior of participants and the nature of the data being processed.

To address these challenges, future research should focus on several key areas. First, there is a critical need for more extensive simulation experiments against malicious attacks to validate the effectiveness of privacy techniques. Such experiments should consider various attack vectors and evaluate the robustness of different defense mechanisms under realistic conditions. By simulating different attack scenarios, researchers can identify potential weaknesses in current defenses and develop more effective strategies to mitigate these threats.

Second, the integration of advanced machine learning algorithms for anomaly detection can enhance the resilience of federated learning systems. These algorithms can help identify and respond to anomalous patterns indicative of malicious activity. For instance, employing neural network-based anomaly detection techniques can enable the system to learn complex patterns of normal and malicious behavior, thereby improving the detection and mitigation of sophisticated attacks. Moreover, incorporating unsupervised learning methods could enable the detection of novel attack types that were not previously encountered, further strengthening the system's defense mechanisms.

Third, exploring hybrid approaches that combine multiple privacy-preserving techniques can provide a more robust defense. For example, integrating homomorphic encryption (HE) with differential privacy (DP) and secure multi-party computation (SMC) can leverage the strengths of each method while compensating for their individual weaknesses. This multilayered approach can offer stronger protection against a broader range of attack vectors. Additionally, such hybrid methods can be designed to be adaptive, dynamically selecting and applying the most appropriate combination of techniques based on the current threat landscape and system requirements.

Another area of focus should be the development of real-time monitoring and response systems within federated learning environments. These systems would enable the continuous monitoring of data and model updates for signs of tampering or malicious activity. When a potential threat is detected, the system could automatically trigger countermeasures, such as isolating the compromised client, retraining the model with adjusted weights, or initiating a secure communication protocol to prevent further damage.

Furthermore, improving the transparency and explainability of federated learning models is crucial for enhancing trust and facilitating the detection of adversarial behavior. By making the decision-making processes of these models more interpretable, researchers and practitioners can better understand how and why certain decisions are made, which in turn can help identify and address potential vulnerabilities that attackers might exploit. This transparency is especially important in collaborative environments where multiple stakeholders must ensure that the system operates fairly and securely.

Moreover, the development of more efficient algorithms for key generation and management, as well as improved methods for ciphertext compression, can help address the computational and storage burdens associated with privacy-preserving techniques. Optimizing communication protocols to reduce the bandwidth required for transmitting encrypted data is also crucial for enhancing the scalability and efficiency of federated learning systems. This is particularly important in scenarios where network resources are limited or where latency could significantly impact the system's performance.

In conclusion, while significant progress has been made in developing privacy-preserving techniques for HFedRec, there is still much work to be done to achieve a balanced and robust defense against malicious attacks. By conducting rigorous simulation experiments, integrating advanced anomaly detection algorithms, and exploring hybrid privacy-preserving approaches, the research community can advance toward more secure and trustworthy federated learning environments. Addressing these challenges will be crucial for ensuring that federated learning systems can provide reliable and confidential recommendations, even in the presence of adversarial participants. The ongoing evolution of attack strategies and the complexity of federated learning systems underscore the need for continuous innovation in privacy-preserving techniques, ensuring that these systems remain resilient in the face of emerging threats.

References

1. Abadi M, Chu A, Goodfellow I, McMahan B, Mironov I, Talwar K, Zhang L (2016) Deep learning with differential privacy. In: Proceedings of the 2016 ACM SIGSAC conference on computer and communications security (CCS)
2. Acar A, Aksu H, Uluagac S (2018) A survey on homomorphic encryption schemes: theory and implementation. ACM Comput Surv 51(4):79
3. Adomavicius G, Kwon Y (2021) Privacy-preserving recommendation systems: a survey. In: Proceedings of the 2021 ACM SIGKDD international conference on knowledge discovery and data mining (KDD)
4. Ammad-ud din M, Ivkin N, Ajmal M (2019) Federated matrix factorization: mitigating data heterogeneity and incentivizing participants. In: Proceedings of the 2019 IEEE international conference on big data (big data)
5. Beaver D, Micali S, Rogaway P (1990) How to share a function and how to share a secret. In: Proceedings of the 2018 ACM SIGSAC conference on computer and communications security (CCS)
6. Bellare M, Hoang VT, Rogaway P (2012) Foundations of garbled circuits. In: Proceedings of the 2012 ACM conference on computer and communications security (CCS)
7. Bonawitz K, Ivanov V, Kreuter B (2017) Practical secure aggregation for privacy-preserving machine learning. In: Proceedings of the 2017 ACM SIGSAC conference on computer and communications security (CCS)
8. Bonawitz K, Ivanov V, Kreuter B (2017) Secure aggregation for federated learning. In: Proceedings of the 2017 ACM SIGSAC conference on computer and communications security (CCS)
9. Bost R, Popa RA, Tu S, Goldwasser S (2015) Privacy-preserving k-means clustering with fully homomorphic encryption. In: Proceedings of the 2015 ACM SIGSAC conference on computer and communications security (CCS)
10. Bu Z, Dong J, Kairouz P (2020) Federated learning with rényi differential privacy. In: Proceedings of the 2020 international conference on learning representations (ICLR)
11. Chai L, Sun L, Jin Z (2020) Federated matrix factorization for privacy-preserving recommendation. In: Proceedings of the 2020 IEEE international conference on big data (big data)
12. Chen J, Wang Y (2022) DGREC: differentially private federated graph recommendation via rényi differential privacy. In: Proceedings of the 2022 ACM SIGKDD international conference on knowledge discovery and data mining (KDD)
13. Chen J, Wang Y, Shen Y (2022) ClusterAttack: poisoning federated recommendation systems with clustering-based item embedding perturbation. In: Proceedings of the 2022 ACM SIGIR conference on research and development in information retrieval (SIGIR)
14. Cheon JH, Kim A, Kim M, Song Y (2017) Homomorphic encryption for arithmetic of approximate numbers. In: Proceedings of the 2017 ACM SIGSAC conference on computer and communications security (CCS)
15. Chor B, Goldreich O, Kushilevitz E, Sudan M (1988) Oblivious transfer and polynomial evaluation. In: Proceedings of the 20th annual ACM symposium on theory of computing (STOC)
16. Dwork C, Roth A (2014) Algorithmic foundations of differential privacy. Found Trends Theor Comput Sci 9(3–4):211–407
17. Feldman V, Zrnic T (2021) Rényi differential privacy for machine learning: a tutorial. In: Proceedings of the 2021 IEEE international conference on machine learning (ICML)
18. Feng Y, Du Y, Lin Y (2022) FedPerGNN: privacy-preserving federated learning for graph neural networks. In: Proceedings of the 2022 IEEE international conference on data mining (ICDM)
19. Goldreich O, Micali S, Wigderson A (1987) Garbled circuits for secure multi-party computation. In: Proceedings of the 2018 ACM SIGSAC conference on computer and communications security (CCS)

20. Huang D (2021) A survey on randomized response techniques for privacy-preserving data collection. ACM Comput Surv 54(1):20
21. Huang L, Yao Z, Zhou J (2021) FedMMF: federated multi-view matrix factorization. In: Proceedings of the 2021 ACM international conference on information and knowledge management (CIKM)
22. Huang L, Zhou J, Zheng J (2021) EIFedMF: privacy-preserving federated matrix factorization with homomorphic encryption and oblivious transfer. In: Proceedings of the 2021 ACM SIGIR conference on research and development in information retrieval (SIGIR)
23. Jiang Z, Chen X, Xu J (2022) IMIA: membership inference attack in federated recommendation systems. In: Proceedings of the 2022 ACM SIGKDD international conference on knowledge discovery and data mining (KDD)
24. Jin X, Liu T, Li D (2021) Federated collaborative filtering for privacy preserving personalized recommendation. In: Proceedings of the 2021 ACM international conference on information and knowledge management (CIKM)
25. Kim J, Jeong JH (2020) Privacy-preserving collaborative filtering based on secret sharing. In: Proceedings of the 2020 ACM SIGIR conference on research and development in information retrieval (SIGIR)
26. Kim H, Jeong JH, Kim SW (2020) FCMF: federated collaborative matrix factorization for recommender systems. In: Proceedings of the 2020 ACM SIGKDD international conference on knowledge discovery and data mining (KDD)
27. Kolesnikov V, Schneider T (2008) Efficient garbling from a fixed-key blockcipher. In: Proceedings of the 2018 IEEE symposium on security and privacy (SP)
28. Li B, Liu Y, Cheng W (2021) A survey on attack and defense in federated learning. In: Proceedings of the 2021 IEEE international conference on data engineering (ICDE)
29. Li J, Huang R, Liu Y (2021) FedAttack: malicious attacks on federated learning with data poisoning. In: Proceedings of the 2021 ACM SIGKDD international conference on knowledge discovery and data mining (KDD)
30. Li D, Li Y, Chen L (2022) Federated point-of-interest recommendation. In: Proceedings of the 2022 ACM SIGIR conference on research and development in information retrieval (SIGIR)
31. Lin H, Yang J, Zheng Z (2021) Federated learning with secret sharing. In: Proceedings of the 2021 ACM SIGKDD international conference on knowledge discovery and data mining (KDD)
32. Lin H, Zheng Z, Yang J (2022) FeSoG: federated social graph neural networks for recommendation. In: Proceedings of the 2022 ACM SIGIR conference on research and development in information retrieval (SIGIR)
33. Lindell Y (2017) A survey on secure multiparty computation protocols. In: Proceedings of the 2017 IEEE symposium on security and privacy (SP)
34. Liu Y, Zhang L, Tang J (2016) Perturbation amplification and its impact on differentially private data analysis. In: Proceedings of the 2016 ACM SIGSAC conference on computer and communications security (CCS)
35. Liu Y, Yang Q, Xu Y (2020) Achieving differential privacy in federated learning with rényi differential privacy. In: Proceedings of the 2020 IEEE international conference on big data (big data)
36. Liu H, Jin X, Liu T (2021) Fast adapting differentially private stochastic gradient descent for federated recommendation. In: Proceedings of the 2021 ACM SIGKDD international conference on knowledge discovery and data mining (KDD)
37. Liu H, Liu T, Tang J (2021) Efficient federated learning for recommendation systems with secured model aggregation. In: Proceedings of the 2021 ACM SIGKDD international conference on knowledge discovery and data mining (KDD)
38. Liu T, Liu H, Tang J (2022) FedRecAttack: attacking federated recommendation algorithms with model poisoning. In: Proceedings of the 2022 ACM SIGIR conference on research and development in information retrieval (SIGIR)
39. McMahan HB, Ramage D, Talwar K, Zhang L (2018) Learning differentially private recurrent language models. In: Proceedings of the 2018 international conference on learning representations (ICLR)

40. Mironov I (2017) Rényi differential privacy. In: Proceedings of the 2017 IEEE symposium on security and privacy (SP)
41. Narayanan A, Shmatikov V (2008) Robust de-anonymization of large sparse datasets. In: Proceedings of the 29th IEEE symposium on security and privacy (SP)
42. Paillier P (1999) Public-key cryptosystems based on composite degree residuosity classes. In: Proceedings of the 17th annual international conference on the theory and applications of cryptographic techniques (EUROCRYPT)
43. Qi Y, Li Y, Cui B (2020) FedNewsRec: privacy-preserving news recommendation with federated learning. In: Proceedings of the 2020 international conference on information and knowledge management (CIKM)
44. Rong Q, Liu J, He X (2022) A-ra: attacking federated recommendation with random approximation. In: Proceedings of the 2022 ACM international conference on information and knowledge management (CIKM)
45. Shamir A (1979) How to share a secret. Commun ACM 22(11):612–613
46. Shokri R, Stronati M, Song C (2017) A survey on membership inference attacks in machine learning. In: Proceedings of the 2017 ACM SIGSAC conference on computer and communications security (CCS)
47. Stinson DR (2018) A survey on secret sharing schemes. In: Proceedings of the 2018 IEEE international symposium on information theory (ISIT)
48. Sun Z, Yin H, Wang Y (2022) Differentially private federated recommendation with social connections. In: Proceedings of the 2022 ACM SIGKDD international conference on knowledge discovery and data mining (KDD)
49. Truex S, Baracaldo N, Anwar A (2019) Secure aggregation for machine learning in federated settings. In: Proceedings of the 2019 ACM SIGSAC conference on computer and communications security (CCS)
50. Wang W, Zhang J, Liu J (2021) Federated matrix factorization with secret sharing. In: Proceedings of the 2021 IEEE international conference on big data (big data)
51. Wang Y, Zhang W, Shen Y (2022) FedSeqRec: federated sequence recommendation with heterogeneous data. In: Proceedings of the 2022 ACM SIGKDD international conference on knowledge discovery and data mining (KDD)
52. Wang Y, Zhang W, Shen Y (2022) PSMU: a poisoning strategy for model updates in federated recommendation systems. In: Proceedings of the 2022 ACM SIGKDD international conference on knowledge discovery and data mining (KDD)
53. Wang S, Zhang X, Xu T (2022) FedHGNN: privacy-preserving federated heterogeneous graph neural networks. In: Proceedings of the 2022 ACM SIGKDD international conference on knowledge discovery and data mining (KDD)
54. Weinsberg U, Ye M, Zheng J, Yolum P, de Veciana G (2012) Blur Me! inferring and obfuscating user gender based on ratings. In: Proceedings of the 2012 ACM international conference on web search and data mining (WSDM)
55. Wu J, Zhang S, Qin Z (2020) FLASHE: a federated learning framework for secure and efficient aggregation of heterogeneous data. In: Proceedings of the 2020 IEEE international conference on big data (Big Data)
56. Xiang T, Li Y, Xu C (2021) Federated neural collaborative filtering for recommendation with privacy-preserving representation learning. In: Proceedings of the 2021 international joint conference on neural networks (IJCNN)
57. Xie G, Huang H, Wang B (2020) HN3S: a privacy-preserving framework for collaborative filtering using secret sharing and homomorphic encryption. IEEE Trans Knowl Data Eng 32(6):1150–1163
58. Xu H, Yin H, Li T (2021) KG-FedTrans4REC: knowledge graph enhanced federated transfer learning for privacy-preserving recommendation. In: Proceedings of the 2021 ACM SIGKDD international conference on knowledge discovery and data mining (KDD)
59. Yao ACC (1982) Protocols for secure computations. In: Proceedings of the 23rd annual IEEE symposium on foundations of computer science (FOCS)

60. Yao Z, Huang L, Zhou J (2021) FedGME: federated graph-based matrix embedding for collaborative filtering. In: Proceedings of the 2021 ACM international conference on information and knowledge management (CIKM)
61. Ye T, Zhu X, Guo H (2020) Towards robust randomized response for privacy-preserving data analysis. In: Proceedings of the 2020 IEEE international conference on data mining (ICDM)
62. Zahur S, Rosulek M, Evans D (2015) Faster garbling of circuits and applications to shorter rounds in secure computation. In: Proceedings of the 2018 ACM SIGSAC conference on computer and communications security (CCS)
63. Zhang W, Zhang S, Cai Z, Yu S, Gu H, Qin Z (2013) PrivCheck: privacy-preserving check-in data publishing for personalized location based services. In: Proceedings of the 2013 IEEE international conference on computer communications (INFOCOM)
64. Zhang Z, Lu D, Liang S, Shen X, Zhang J (2019) On gradient clipping in deep learning and how to utilize it. In: Proceedings of the 2019 international conference on machine learning (ICML)
65. Zhang L, Lin H, Wu J (2022) Federated neural collaborative filtering with privacy-preserving and model agnostic aggregation. Knowl-Based Syst 239:107946
66. Zhang W, Liu T, Yang B (2022) PipAttack: poisoning federated recommender systems with item promotion. In: Proceedings of the 2022 ACM SIGIR conference on research and development in information retrieval (SIGIR)
67. Zhao S, Zheng L, Gao F (2020) A survey on privacy-preserving federated learning: from methods to applications. IEEE Trans Big Data
68. Zheng Y, Capra L, Wolfson O, Yang H (2014) Urban computing: concepts, methodologies, and applications. ACM Trans Intel Syst Technol 5(3):38
69. Zhou R, Zhang Y (2018) Degree-preserving randomized response for graph privacy. In: Proceedings of the 2018 IEEE international conference on data engineering (ICDE)
70. Zhou R, Luo P, Zheng J (2021) Finding: federated inductive learning via nearest neighbor clustering. In: Proceedings of the 2021 ACM international conference on information and knowledge management (CIKM)
71. Zhu H, Xu W, Zhang F (2021) Semi-supervised deep federated learning for edge computing. IEEE Trans Mobile Comput 20(6):1808–1821

Chapter 4
Federated Issues in Cross-Device Federated Recommendation

Abstract This chapter explores two critical aspects of federated learning: communication optimization and fairness perception. It introduces communication-efficient techniques, such as knowledge distillation, parameter compression, edge computing, and client clustering, to address resource constraints and enhance model convergence. For fairness, methods like reweighting parameter aggregation, regularization, multi-task learning, and differential privacy are discussed to mitigate biases and promote equity. The chapter highlights how these strategies improve efficiency and fairness in federated learning for IoE scenarios and outlines challenges for future research, including dynamic data heterogeneity and balancing fairness with performance.

4.1 Communication Optimization

In the context of federated recommendation systems (FeDRS), communication optimization plays a critical role, especially in cross-device scenarios where data is horizontally partitioned across a vast number of devices. Horizontal federated recommendation (HFedRec) is particularly suited for this setting, as each client holds data from different users but shares common features. Due to the decentralized nature of this setup, communication between devices and the central server becomes a bottleneck. Efficient communication is essential to maintain model performance while minimizing delays and reducing bandwidth usage, making it a key focus for HFedRec. This section explores techniques that are specifically designed to address these challenges, optimizing the communication process in cross-device federated environments. While some concepts may be transferable to other types of federated recommendation systems, the solutions discussed here are tailored to the unique demands of HFedRec.

The ever-growing scale of users and items contributes to the sparsity issue of RecSys [3], meaning that a portion of user-item interactions are insufficient to cover the broad user groups and candidate items. To address this challenge and sustain recommendation performance, it is imperative to involve more clients in the

X. Kong et al., *Cross-device Federated Recommendation*, Machine Learning: Foundations, Methodologies, and Applications,
https://doi.org/10.1007/978-981-96-3212-1_4

federation. However, the inclusion of resource-constrained clients (e.g., those with low bandwidth) and the reliance on a single-point server can introduce significant bottlenecks, ultimately hindering the acceleration of federated model convergence.

In this section, we concentrate on communication-efficient techniques, as shown in Table 4.1. Unlike the specific models listed in Table 3.1, these techniques are more general and not inherently tied to any particular neural network structure. This versatility allows them to be applied across various federated learning scenarios, making them crucial for optimizing communication and ensuring the efficient operation of federated systems as they scale.

4.1.1 Knowledge Distillation

Knowledge distillation is a powerful technique in machine learning, enabling a smaller *student* model to leverage the supervision provided by a larger, more complex *teacher* model to achieve comparable or superior performance [12]. The "knowledge" transferred from the teacher to the student can take multiple forms, primarily categorized into two types: *feature-based knowledge*, where hidden layers' features are transferred, and *response-based knowledge*, where soft logits inferred by the final prediction layer are transferred. For the latter, given a user u_i and the items V_{u_i} that the user has interacted with, the distillation loss is typically formulated as follows:

$$l_{distill} = \mathrm{KL}\left(\mathrm{Logit}_t(u_i, V_{u_i}|\theta_t), \mathrm{Logit}_s(u_i, V_{u_i}|\theta_s)\right) \tag{4.1}$$

where $\mathrm{KL}(\cdot)$ denotes the Kullback-Leibler divergence and $\mathrm{Logit}_t(\cdot)$ and $\mathrm{Logit}_s(\cdot)$ represent the logits predicted by the teacher and student models, respectively. This loss function encourages the student model to mimic the output distribution of the teacher model, allowing it to capture the teacher's complex decision-making processes while remaining smaller and more computationally efficient.

In federated learning, the advantages of knowledge distillation become particularly evident. Federated systems often require models that can perform well while being lightweight enough to minimize communication costs and computational burdens on client devices. Knowledge distillation serves this purpose by enabling a smaller student model to inherit the rich information learned by a larger teacher model, ensuring high performance without compromising efficiency. This trade-off is crucial in federated learning, where both communication and computational resources are often limited.

An exemplary application of knowledge distillation in federated learning is FedKD [21]. In this framework, a large teacher model is maintained on each client and is distilled by a smaller global student model in a mutual-distillation process. The student model, which has a significantly smaller parameter size, is the only model globally aggregated and shared among clients, substantially reducing communication costs. Furthermore, distillation losses are computed adaptively

Table 4.1 Representative works of communication-efficient HFedRec

Publications	Type	U/O	Tech	Tar	Publications	Type	U/O	Tech	Tar
FedKD [21]	D	O	KD[a], Comp[b]	Model[a], Param[b]	URFL [23]	D	U	Edge	Distr
ECCT [41]	D	O	KD	Model	LightFR [5]	M	U	Comp	Model, Param
JointRec [39]	D	U	Edge[a], Comp[b]	Distr[a], Param[b]	DGREC	D	U	Comp	Param
HFCCD [27]	M	U	Edge[a], Comp[b]	Distr[a], Param[b]	CF-FedSR [44]	D	U	Cluster	ItemSequence
HFSA [7]	D	U	Edge	Distr	FedFast [20]	D	U	Cluster	UserMeta
CCF-CRR [47]	D	U	Edge	Distr	PerFedRec [34]	D	U	Cluster	UserEmbed
GIMA [42]	D	U	Edge	Distr	SemiDFEGL [40]	D	U	Cluster	Graph, ItemEmbed

Type: M, statistical machine learning; D, deep learning. **U/O:** U, cross-user federation; O, cross-organization federation

Tech (communication-efficient techniques)**:** KD, knowledge distillation [15]; Comp, parameter compression; Edge, edge computing [32]; Cluster, cluster-based client sampling

Tar (the applied targets of Tech)**:** Model, federated model; Param, transferred parameters; Distr, distribution architecture; UserEmbed, user embedding; ItemEmbed, item embedding; ItemSequence, encoded item sequences; Graph, encoded local graphs; UserMeta, user metadata

[a,b]Indicate which target a communication-efficient technique is applied to

based on the descent degree of the loss, dynamically balancing the learning process between the teacher and student models. This adaptive approach ensures that the system maintains high performance despite evolving model parameters and varying data distributions across clients.

In contrast, ECCT [41] introduces a server-based distillation strategy where the teacher model is maintained centrally. In this setup, clients retain only small student models, and prediction logits are transmitted back to the server to optimize the teacher model, thereby reducing the computational and storage burden on the client devices. This strategy is particularly advantageous in scenarios where clients have limited computational resources but the server has ample capacity to handle more complex models. By centralizing the teacher model, the system ensures that clients are only required to manage lightweight models, thereby alleviating both computational and communication burdens.

One of the key benefits of knowledge distillation in federated learning is its ability to reduce communication overhead by ensuring that only smaller models or less complex data, such as logits, are transmitted between clients and the server. This is particularly critical in bandwidth-constrained environments typical of federated learning systems. Additionally, by reducing the size of the models that need to be transmitted, knowledge distillation helps mitigate the impact of network latency and bandwidth limitations, ensuring that the system can scale effectively across large numbers of devices. Beyond communication efficiency, knowledge distillation also plays a vital role in managing the computational load on client devices. By transferring knowledge from a large teacher model to a smaller student model, it allows client devices to maintain high performance without the need for extensive local computation, thus making the system more accessible for a diverse range of devices with varying computational capacities.

Moreover, knowledge distillation offers substantial flexibility in deployment. In server-heavy environments, server-based distillation approaches like ECCT can maximize the use of centralized resources while minimizing client-side demands. In contrast, when clients have relatively strong computational capabilities, methods like FedKD can be deployed, allowing for a more decentralized approach that still leverages the advantages of knowledge distillation. This flexibility makes knowledge distillation an ideal solution for federated learning systems deployed in heterogeneous environments, where both communication and computational resources may vary significantly across devices (Fig. 4.1).

In summary, knowledge distillation provides a robust mechanism for enhancing the efficiency and performance of federated learning systems. Techniques like FedKD and ECCT exemplify how knowledge distillation can reduce communication costs, optimize resource utilization, and maintain high model performance, even in resource-constrained federated environments. As federated learning continues to grow in importance across various applications, the role of knowledge distillation in balancing performance, resource usage, and communication efficiency will be critical for the development of scalable and high-performing systems.

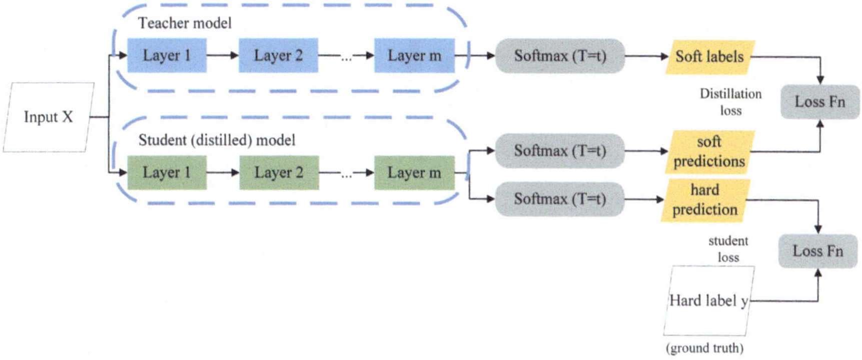

Fig. 4.1 Illustration of knowledge distillation

4.1.2 Parameter Compression

The presence of massive user-item interactions can result in an increasing model size, inevitably posing a challenge for resource-constrained clients in handling heavy communication and intensive computing [18]. The sheer volume of data generated in federated learning environments can overwhelm both the communication channels and the computational capabilities of client devices, particularly those with limited resources. To address these challenges, parameter compression is intended to achieve two primary objectives: **(i)** decrease uplink and downlink burden during client-server transmission, thereby reducing the communication overhead, and **(ii)** enhance computational efficiency in both training and inference phases, allowing client devices to process data more effectively without being bogged down by the complexity of the operations involved.

LightFR [5] is a noteworthy work that satisfies these two intents. It binarizes continuous user/item embeddings into the discrete Hamming space by learning to hash [36]. This transformation involves converting real-valued vectors into binary codes, significantly reducing the data size and simplifying the computational processes. Local discrete optimization and global discrete aggregation modules are tailored to update binary user vectors and the item matrix. The discrete Hamming space significantly reduces computing complexity and communication consumption by up to eight times compared to the Euclidean space, making it highly efficient for federated learning scenarios. This reduction in complexity is particularly beneficial for devices with limited computational power, as it allows them to participate in the federated learning process without being overwhelmed by the computational demands. The efficiency gains from LightFR are crucial in enabling scalable federated learning systems where communication and computation resources are at a premium (Fig. 4.2).

FedKD [21] employs singular value decomposition (SVD) to factorize transmitted gradients of the student model into three smaller matrices. This factorization

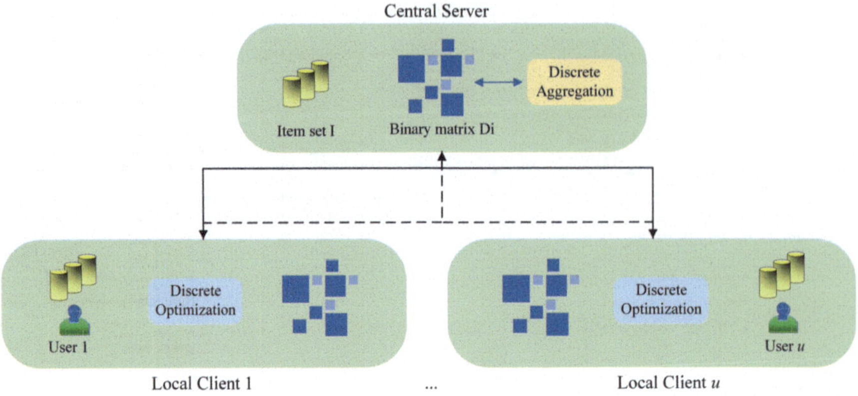

Fig. 4.2 LightFR

reduces the size of the data that needs to be transmitted, thus lowering the communication overhead. The SVD approach not only compresses the data but also retains the essential features needed for effective model training, ensuring that the compressed data is still useful for learning. By breaking down the gradient matrix into lower-dimensional representations, SVD effectively reduces the amount of information that needs to be sent while preserving the critical components necessary for model updates. This approach is particularly advantageous in federated learning settings where minimizing data transmission is crucial for maintaining system efficiency.

Additionally, JointRec [39] combines low-rank matrix factorization (MF) with 8-bit quantization to compress uplink model weights. This approach not only reduces the data volume but also maintains the model's performance by preserving essential information in a compressed form. The use of 8-bit quantization is particularly advantageous as it strikes a balance between compression efficiency and the preservation of model accuracy. By quantizing the model weights to lower bit precision, JointRec significantly reduces the size of the data that needs to be transmitted while still ensuring that the compressed weights can effectively contribute to the model's learning process. This balance is critical in scenarios where communication bandwidth is limited, and maintaining model accuracy is paramount.

HFCCD [27] utilizes count sketch to map high-dimensional gradients to a fixed-size sketch. This technique effectively reduces the dimensionality of the data, making it more manageable for transmission and processing. Count sketch is a probabilistic data structure that provides a compact representation of the input data, allowing for significant compression without substantial loss of information. The compact sketches generated by this method can be transmitted efficiently, reducing the communication load on the network. Moreover, HFCCD also refers to the design of the 1-bit encoder [31], which encodes each element of the transmitted gradients as -1 or 1. This binary encoding minimizes the communication cost and simultaneously satisfies Rényi differential privacy, providing a robust privacy guarantee while reducing the data size. The 1-bit encoding approach is highly efficient in terms of both computation and communication, making it suitable for large-scale federated learning systems where privacy and efficiency are both critical considerations.

The applied compression schemes in [21, 27, 39] also require decompression or decoding before parameter aggregation or other operations. These steps are essential to reconstruct the original information from the compressed data, ensuring that the integrity and performance of the federated learning model are maintained. The decompression process involves reversing the compression transformations, such as matrix multiplication in the case of SVD, to restore the data to a form that can be used for further learning and aggregation. This step is crucial for maintaining the accuracy and effectiveness of the model, as it ensures that the information lost during compression is adequately recovered during the decompression phase.

In summary, parameter compression is a critical technique for enhancing the efficiency of federated learning systems, particularly in scenarios involving resource-constrained clients. By reducing the data volume for transmission and optimizing computational processes, these methods address the challenges of communication overhead and computational load. The continued development and refinement of compression techniques will be crucial for enabling scalable and efficient federated learning, ensuring that even clients with limited resources can participate effectively. The evolution of these techniques will likely involve more sophisticated algorithms that can further compress data without losing critical information, as well as improved methods for balancing compression efficiency with model accuracy. As federated learning continues to expand into various applications and domains, the role of parameter compression in maintaining system scalability and performance will become increasingly important.

4.1.3 Edge Computing

The Internet of Everything (IoE) era is continually driving the integration of terminal devices to unlock unprecedented value beyond their individual capabilities, which is equally observable in the federated distribution environment [2]. The rapid expansion of connected devices has led to an exponential increase in data generation and processing demands. In a federated learning context, this growth presents significant challenges for traditional centralized servers, which can become bottlenecks due to their limited bandwidth and computational capacities. It is intractable for a single server to ensure the immediacy of recommendation services when connected devices gradually approach its bandwidth and computing limits. As the number of participating devices continues to grow, the need for more efficient and scalable computational infrastructures becomes paramount. This is where edge computing emerges as a crucial technology, offering a decentralized approach to data processing that significantly enhances the performance and scalability of federated learning systems.

With edge servers deployed closer to end users, edge computing can expand the cloud infrastructure and realize low-latency communication [32, 39]. By processing data closer to its source, edge computing reduces the distance that data must travel, thereby decreasing latency and improving response times. This proximity to end users allows for faster and more efficient processing, which is particularly beneficial for applications requiring real-time data analysis, such as recommendation systems in federated learning. The ability to quickly process and analyze data at the edge also helps to reduce the load on central servers, freeing up resources and enabling the system to handle a larger number of clients more effectively.

One of the dominant hierarchical architectures is cloud-edge-end collaboration [37]. In this architecture, the central server, edge servers, and end devices work together to distribute the computational load and optimize the overall system performance. As a bridge between the central server and clients, edge clusters aim to divert long-distance communication consumption and handle partial data processing. This hierarchical approach ensures that data processing is distributed effectively across the network, reducing the load on central servers and improving the efficiency of the entire system. By offloading certain tasks to edge servers, the system can better manage the computational and communication demands, ensuring that recommendation services remain responsive and efficient even as the number of connected devices increases.

HFSA [7] proposes a semi-asynchronous aggregation mechanism and decides the cycle of edge-end collaboration by local training efficiency. This method allows the system to adapt to the varying computational capabilities and network conditions of different clients. For example, clients too slow to catch up with others' progress are only permitted to receive data from edge servers. This approach ensures that the slowest clients do not bottleneck the entire system, allowing more efficient utilization of resources. By dynamically adjusting the collaboration cycles between edge and end devices, HFSA optimizes the balance between computational load and communication efficiency, ensuring that all clients can contribute to the federated learning process without overwhelming the system.

In addition to handling computational load, edge servers play a crucial role in data aggregation and preprocessing. By performing these tasks closer to the data source, edge servers can reduce the amount of data that needs to be transmitted to the central server, further decreasing communication latency and bandwidth usage. This proximity also allows for more immediate feedback and updates, which is critical for real-time applications. The ability to quickly aggregate and preprocess data at the edge enables the system to provide timely and relevant recommendations, enhancing the overall user experience.

CCF-CRR [47] probes into the cruising route recommendation for taxis and passengers. In this scenario, an edge server is responsible for training a federated recommendation model with taxis in the same region. By receiving predicted passenger waiting times from edge servers, the cloud server can ensure the taxi demand-supply equilibrium among regions. This method leverages the proximity of edge servers to provide more timely and region-specific recommendations, enhancing overall service efficiency. The use of edge servers in this context allows for dynamic adjustments based on real-time data, improving the accuracy and relevance of the recommendations. This approach not only optimizes resource allocation among taxis but also improves the quality of service for passengers, demonstrating the potential of edge computing in location-based and real-time recommendation scenarios.

HFCCD [27] introduces a compression scheme and differential privacy to improve the reliability and confidentiality of multi-party collaboration. The combination of compression and privacy-preserving techniques ensures that data can be securely and efficiently processed across different parties in the federated learning framework. Compression techniques help in reducing the data size, making transmission faster and less resource-intensive, while differential privacy ensures that individual data points remain confidential, protecting user privacy even in a distributed environment. This dual focus on efficiency and privacy is crucial for federated learning systems that must balance the need for fast, scalable processing with the imperative to protect user data.

Mobile edge computing (MEC) [52] is another distributed architecture where computing, storage, and network resources are strategically assigned closer to mobile terminals or edge devices. MEC brings computation and data storage closer to the location where it is needed to improve response times and save bandwidth. By offloading data processing to edge devices, MEC reduces the burden on the central server and minimizes the latency involved in transmitting data over long distances. This architecture is particularly beneficial for applications requiring real-time data processing and low-latency responses, such as augmented reality, autonomous driving, and IoT-based services. MEC enhances the overall user experience by providing faster and more reliable services, which is critical in today's fast-paced digital environment. The ability to process data at the edge also enables more sophisticated and personalized services, as the system can respond more quickly to individual user needs and preferences (Fig. 4.3).

For inspiring users to participate in MEC-based HFedRec, GIMA [42] proposes an incentive mechanism based on the two-stage Stackelberg game [46]. The profit-oriented competition among users consequently accelerates model convergence. This game-theoretic approach incentivizes users to contribute their data and computational resources by offering rewards, thus enhancing the overall performance and speed of the federated learning process. By aligning the incentives of users with the goals of the federated learning system, GIMA ensures active participation and cooperation, which are essential for the success of distributed learning models. This approach not only improves the quality and speed of the learning process but also fosters a more collaborative and motivated user base, which is crucial for the scalability and sustainability of federated learning systems.

Due to the spatiotemporal varying subjectivity (e.g., interests) of mobile users, URFL [23] devises an unsupervised recurrent federated learning approach to predict global content popularity. It sacrifices a tolerable rise in data traffic within the MEC network while improving recommendation accuracy. By accounting for the

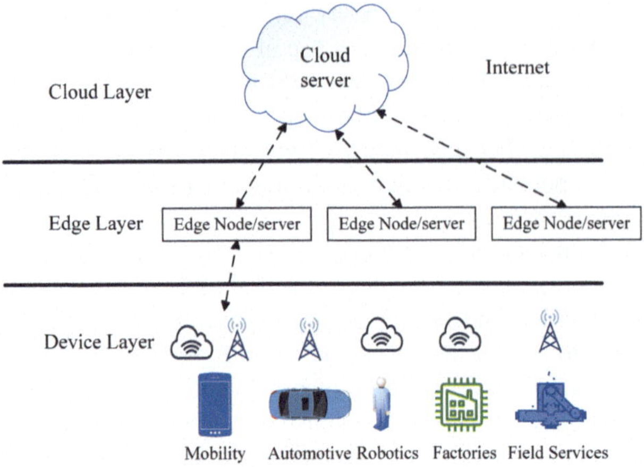

Fig. 4.3 Edge computing

dynamic nature of user preferences and the temporal aspects of data, this method enhances the relevance and quality of the recommendations provided to users. The ability to adapt to changing user behaviors and preferences in real time ensures that the recommendation system remains effective and user-centric. This dynamic adaptation is particularly important in mobile and IoT environments, where user contexts and preferences can change rapidly, necessitating a responsive and flexible recommendation system.

Edge computing, particularly through MEC, plays a crucial role in optimizing federated learning by bringing computational resources closer to the data source. This reduces latency, improves response times, and ensures more efficient use of network resources. By leveraging the strengths of edge computing, federated learning systems can achieve higher performance and scalability, making them more suitable for real-time and resource-intensive applications. The integration of edge computing into federated learning frameworks not only enhances the system's ability to handle large-scale data processing but also ensures that the recommendations generated are timely, relevant, and responsive to user needs.

In summary, edge computing, particularly through MEC, plays a crucial role in optimizing federated learning by bringing computational resources closer to the data source. This reduces latency, improves response times, and ensures more efficient use of network resources. As federated learning continues to evolve, the integration of edge computing techniques will be essential for developing scalable, efficient, and real-time recommendation systems. The continued advancement of edge computing technologies will further enhance the capabilities of federated learning, enabling it to meet the growing demands of modern applications and services. By combining the strengths of edge computing with the collaborative nature of federated learning, future systems will be able to deliver superior performance, scalability, and user satisfaction.

4.1.4 Client Clustering

Different user interests are not independent and identically distributed (non-IID) among clients, such that random client sampling before a communication round may neglect co-occurrence preference patterns of certain user groups. In federated learning, this non-IID nature poses significant challenges to model convergence and performance, as the aggregated global model may not adequately capture the diverse and specific preferences present across different client groups. The objective of client clustering is to proportionally sample representative clients (i.e., delegates) from each community, thus reducing intra-cluster non-IIDness and accelerating the federated training process. By organizing clients into clusters based on similarity in their data distributions or behavioral patterns, federated learning systems can achieve more efficient communication, faster convergence, and improved personalization in recommendations.

FedFast [20] is a pioneering approach that implements client clustering by utilizing summary statistics of user profiles. It strategically groups clients with similar user profiles and selects delegates from each cluster to participate in the current communication round. This targeted selection ensures that the aggregated model updates are representative of the diverse user preferences across different clusters. The aggregation strategy in FedFast involves sharing the trained weights of delegates with other clients within the same cluster, effectively disseminating learned knowledge throughout the group. This method significantly enhances convergence speed; experiments demonstrate that non-clustered aggregation requires 94% more communication rounds to match the performance achieved by FedFast. The efficiency gains from this approach are substantial, reducing both communication overhead and computational resources required for model training. Additionally, by focusing on cluster-specific updates, FedFast enhances the personalization of recommendations, catering more accurately to the preferences of users within each cluster.

Similarly, Cali3F [26] applies the principles of FedFast to develop a fairness-aware hierarchical federated learning framework for recommendation systems. By incorporating client clustering, Cali3F ensures that model training is both efficient and equitable across different user groups. This framework addresses potential biases by guaranteeing that minority or underrepresented groups are fairly included in the training process, thus producing recommendation models that are more inclusive and balanced. The clustering mechanism helps in identifying and grouping users with similar characteristics, enabling the model to learn from diverse data sources effectively and mitigating the dominance of majority groups in the training process. This leads to fairer outcomes and improves user satisfaction across the board.

CF-FedSR [44] advances the client clustering paradigm by leveraging the temporal dynamics of user behavior. It clusters concatenated vectors of short- and long-term preferences encoded by GRU4Rec [14], capturing both immediate and evolving user interests over time. This method acknowledges that user preferences are dynamic and context-dependent, changing with time and circumstances. By incorporating both short-term and long-term behavioral data into the clustering process, CF-FedSR can form more accurate and meaningful clusters, leading to recommendations that are timely and contextually relevant. This temporal sensitivity enhances the model's adaptability and responsiveness to shifts in user behavior, improving overall recommendation accuracy and user engagement.

PerFedRec [34] and PerFedRec++ [33] utilize user embeddings to perform clustering, effectively capturing the latent features and complex relationships underlying user preferences. User embeddings are low-dimensional representations that encapsulate various aspects of user behavior and interests. By clustering these embeddings, the models can identify groups of users with similar latent preferences, facilitating more targeted and personalized recommendations. This approach also allows for efficient handling of high-dimensional and sparse data common in recommendation systems. The enhanced clustering based on user embeddings leads to better model generalization and robustness, as it captures subtle and intricate patterns in user behavior that may not be apparent through superficial statistical measures.

Apart from server-driven sampling of representative clients, SemiDFEGL [40] introduces a novel paradigm where a client is empowered to communicate directly with other clients within the same cluster. This peer-to-peer (P2P) collaboration facilitates cross-client feature diffusion and decreases the number of heavy cloud-end communications, thereby reducing latency and improving scalability. By enabling direct interactions among clients, SemiDFEGL allows for more localized and efficient knowledge sharing, harnessing the collective intelligence of the cluster without overburdening the central server. This decentralized approach enhances data privacy, as sensitive information is shared within a confined group, and promotes faster convergence by exploiting the proximity and similarity among clients.

Cluster-based federation offers numerous benefits beyond efficient training. EIFedMF demonstrates that clustering can enhance privacy preservation by limiting the exposure of sensitive data to a smaller, more controlled group. By restricting data sharing and model updates within clusters, the risk of data leakage and privacy breaches is significantly reduced. This is particularly important in scenarios where data privacy is a major concern, as it ensures compliance with regulations and maintains user trust. Additionally, overall recommendation performance is improved, as evidenced in studies like ReFRS [22], CdFed [51], CPF-POI [48], and PrefFedPOI [16]. These works showcase that by aligning user groups with similar preferences, recommendation models can achieve higher accuracy and relevance. The clustered approach ensures that recommendations are more tailored to the specific needs and preferences of distinct user segments, enhancing user satisfaction and engagement.

Addressing the cold-start issue is another significant advantage of client clustering. FINDING employs a clustering strategy to effectively handle new users or items with limited interaction data. By associating these newcomers with established clusters based on available minimal information, the system can infer preferences from similar users or items within the cluster, thus mitigating the cold-start problem.

This approach enables the recommendation system to provide relevant suggestions to new users or for new items from the outset, improving the user experience and facilitating quicker integration into the system's ecosystem.

Moreover, client clustering enhances scalability in federated learning systems. As the number of clients grows, clustering helps in managing the increased complexity by organizing clients into manageable and coherent groups. This hierarchical structuring simplifies coordination and aggregation of updates, ensuring that the system remains efficient and responsive even as it scales to accommodate vast numbers of users. Effective clustering strategies also enable selective participation in training rounds, optimizing resource allocation and reducing unnecessary communication and computation.

In summary, client clustering is a powerful technique for enhancing the efficiency and effectiveness of federated learning in recommendation systems. By organizing clients into meaningful clusters based on similarities in data distributions, behavioral patterns, or latent preferences, models can better capture user-specific characteristics, reduce communication overhead, and improve overall performance. The ability to manage non-IID data, preserve privacy, address the cold-start problem, and enhance scalability makes client clustering an indispensable strategy in federated learning. Continued exploration and refinement of clustering methods, including adaptive and dynamic clustering techniques, will be crucial for advancing federated learning and addressing the diverse challenges of personalized recommendation systems in increasingly complex and data-rich environments.

4.1.5 Summary and Challenges

The communication-efficient techniques discussed above exchange a diversity of viewpoints on model size, transmission, distributed architecture, and client sampling. These techniques are critical for optimizing communication costs in federated learning environments where the sheer volume of data and the need for frequent updates can strain network resources. In addition to these, special learning paradigms also significantly reduce communication costs. For instance, meta-learning-based FedIS [4] and FedMeta [6] require fewer convergence epochs and less transmission quantity than vanilla aggregation, thereby optimizing communication efficiency. By rapidly adapting to new tasks with minimal data, meta-learning approaches can significantly cut down the communication rounds needed to achieve satisfactory model performance.

Reinforcement learning-based FCF-BTS [30] optimizes the downlink payload by using a bandit model in the server, dynamically adjusting the transmission strategy based on current network conditions. This approach allows the system to make real-time decisions on data transmission, ensuring that the communication load is balanced according to the available bandwidth and network latency. Such dynamic adjustment is crucial in heterogeneous network environments where different clients may have varying levels of connectivity and computational capacity.

EfficientFedRec [43] places single-user models locally, utilizing split learning techniques to further reduce communication overhead. Split learning divides the model between the client and the server, ensuring that only part of the model needs to be trained and transmitted at any given time. This reduces the amount of data exchanged during training and can significantly lower communication costs, especially when dealing with large-scale models or when operating in environments with limited bandwidth. The details of these approaches have been described in Chap. 2. From our point of view, the federated communication issue should be emphasized in real IoE scenarios, since HFedRec is naturally oriented toward a wide range of geographically distributed end devices [24]. This geographical distribution introduces unique challenges in ensuring efficient and reliable communication among the devices involved, as different regions may have different network infrastructures and levels of connectivity.

Therefore, one of the primary challenges is how to incorporate load balancing, service discovery, and fast adaptation in edge computing-empowered HFedRec [17, 53]. Load balancing ensures that computational tasks are evenly distributed across the network, preventing any single device or server from becoming a bottleneck. This is particularly important in federated learning, where tasks such as model training and data aggregation can be computationally intensive. Service discovery allows devices to efficiently locate and communicate with each other, optimizing the overall network performance. This is crucial in dynamic environments where devices may frequently join or leave the network. Fast adaptation ensures that the system can quickly respond to changes in the network environment, such as varying network conditions or the addition/removal of devices. This adaptability is key to maintaining the performance and reliability of federated learning systems in real-world IoE scenarios.

In the future, considerations around electric power, energy utilization, storage, and other hardware resources will be crucial in bringing HFedRec into everyday use. As federated learning systems become more widespread, managing these resources efficiently will become increasingly important. Effective management of these resources can enhance the sustainability and efficiency of federated learning systems. For instance, optimizing energy usage can prolong the operational life of battery-powered devices, which are common in IoT environments. Energy-efficient algorithms and hardware can significantly reduce the power consumption of federated learning tasks, making the system more sustainable. Efficient storage management can ensure that data is available when needed without overwhelming the system. This involves not only optimizing how data is stored but also developing techniques for data compression and deduplication to minimize storage requirements.

Moreover, advancements in hardware technologies can further support the deployment of HFedRec. Innovations in low-power processors, high-speed networking components, and scalable storage solutions can provide the necessary infrastructure to support large-scale federated learning deployments. Low-power processors can perform complex computations while consuming minimal energy, which is essential for battery-operated devices in IoT networks. High-speed networking components can improve the speed and reliability of data transmission, reducing latency and ensuring that models can be trained and updated more quickly. Scalable storage solutions can accommodate the large volumes of data generated by federated learning systems, ensuring that data can be stored and retrieved efficiently.

By integrating these hardware advancements with sophisticated software algorithms, HFedRec can achieve new levels of performance and efficiency. For example, integrating AI-driven resource management algorithms can optimize how computational and storage resources are allocated, ensuring that the system operates at peak efficiency. Additionally, advancements in edge computing can enhance the capabilities of federated learning by enabling more complex computations to be performed closer to the data source, reducing the need for data to be transmitted over long distances.

In summary, while significant progress has been made in developing communication-efficient techniques for federated learning, several challenges remain. Addressing these challenges will require a holistic approach that considers both software and hardware aspects. By incorporating load balancing, service discovery, fast adaptation, and efficient resource management, HFedRec can be effectively deployed in real-world IoE scenarios, making advanced recommendation systems accessible to a broader audience. The continued development of both hardware and software technologies will be crucial for overcoming these challenges, ensuring that HFedRec can operate efficiently and sustainably in a wide range of environments. As these technologies evolve, they will enable HFedRec to meet the growing demand for personalized, real-time recommendations in increasingly complex and dynamic networks.

4.2 Fairness

Ideally, preference data generated by every individual or group (varying in gender, occupation, age, etc.) should have an equal opportunity to contribute to and benefit from the collaborative training of the federated model. Recommendation behaviors that omit disadvantaged groups or only highlight popular tastes result in unfair actions, characterized by discrimination and prejudice [28]. Although it is confined to the horizontal federated recommendation scenario, there is still no universal fairness definition capable of encapsulating all fairness objectives [45]. This complexity is reflected in the diverse fairness metrics and strategies adopted across different studies.

We classify existing efforts into client and group levels to provide a structured overview of the current landscape of fairness in federated learning. A comprehensive review by Li et al. [19] discusses the various fairness challenges and approaches in federated learning. The review highlights that federated learning systems must ensure fairness across different dimensions, including demographic parity, equal opportunity, and equalized odds. The review also emphasizes the importance of developing fairness-aware algorithms that can adapt to the dynamic and heterogeneous nature of federated learning environments. Some of the key insights from the review include the following:

- **Fairness Constraints:** Implementing constraints during the training process to ensure that the model's predictions do not favor or discriminate against any particular group
- **Post-processing Techniques:** Adjusting the model's predictions after training to correct any unfair biases that may have been learned
- **Fair Representation Learning:** Developing methods to learn representations of data that are invariant to sensitive attributes, ensuring that these attributes do not influence the model's predictions
- **Collaborative Fairness:** Encouraging collaboration between clients to share fairness-related insights and strategies, thus promoting a more equitable learning process across the federated network

4.2.1 Client-Level Fairness

Current client-level fairness is devoted to pursuing uniform accuracy among clients. In federated learning, where clients often have varying data distributions and resources, achieving fairness across all participants is challenging but essential. Given clients C and two federated models f_1 and f_2, f_1 is considered fairer than f_2 when

$$\text{std}\left(\{\text{Eval}(f_1)_c\}\right) < \text{std}\left(\{\text{Eval}(f_2)_c\}\right), c \in C \tag{4.2}$$

where $\text{std}(\cdot)$ denotes the standard deviation of clients' metrics $\text{Eval}(\cdot)$ on test data. The lower the standard deviation, the more uniform the performance across different clients, indicating a fairer distribution of model accuracy. Other uniformity functions, such as cosine similarity and entropy, can also measure the performance distribution [1]. These metrics help in assessing how evenly the model's benefits are distributed among the clients, providing a quantitative basis for evaluating fairness. For a more comprehensive overview of fairness notions, readers can refer to [45].

Client-level fairness aims to ensure that all clients achieve similar levels of performance, regardless of the variability in their local datasets. This is particularly important in federated learning scenarios where data heterogeneity can lead to biased models that favor certain clients over others. Without proper fairness considerations, federated learning systems risk amplifying existing inequalities, with clients possessing larger or higher-quality datasets potentially dominating the model's learning process. Achieving fairness requires addressing biases at both the server-side and client-side levels, ensuring that the model performs equitably across all participants.

For the server-side unfairness, classical aggregation strategies weigh the importance of uploaded parameters by the size of local datasets. This approach, however, can lead to disproportionate influence from clients with larger datasets, skewing the global model toward their specific data distributions. Clients with sufficient local data may bias the global recommendation model toward their preference patterns, exacerbating inequalities among participants. One fair solution is to reweight the parameter aggregation process to account for client-level disparities. FL-GMT [54] collects clients' training losses in each communication round. The larger the training loss, the greater the weight coefficient assigned to that client's parameters. The core idea is that a larger local loss represents poorer local model performance, so more resources should be invested to improve it. This approach ensures that clients with less representative data do not disproportionately influence the global model, promoting a more balanced learning process. By focusing on clients with higher training losses, FL-GMT helps to uplift the performance of underperforming clients, thereby improving overall fairness.

Similarly, CF-FedSR [44] generates weight coefficients for the aggregation stage based on a combination of local performance metrics and data size, ensuring a balanced contribution from all clients. This method accounts for both the quantity and quality of the data, striking a balance between leveraging large datasets and ensuring equitable treatment of clients with smaller or more challenging datasets. By adjusting the aggregation weights dynamically, CF-FedSR enhances the fairness of the federated learning process, ensuring that all clients, regardless of their data characteristics, contribute meaningfully to the global model. This approach mitigates the risks of bias accumulation, where certain clients could disproportionately influence the model due to their data advantages (Fig. 4.4).

To further enhance server-side fairness, adaptive aggregation techniques can be employed. These techniques dynamically adjust the weights assigned to each client's parameters based on their current performance and contribution to the global model. Adaptive aggregation allows the federated learning system to continuously balance the influence of different clients, ensuring that no single client dominates the learning process. This approach also helps in maintaining model robustness across diverse and non-IID data distributions. By tailoring the aggregation process to the evolving characteristics of each client, adaptive techniques ensure that the global model remains fair and inclusive throughout the training process.

For the client-side unfairness, local model parameters are typically substituted with the globally aggregated results, guiding all clients to follow a generalized preference pattern. However, the global update direction may be quite distinct from local ones, potentially leading to suboptimal performance for clients whose data significantly deviates from the global trend. To address this issue, Cali3F [26] introduces a regularization term into the loss function to control the differentiated degree between local and global parameters. This regularization ensures that the global model updates do not disproportionately disadvantage any single client's local model, maintaining fairness across the board. Specifically, the regularization term can be defined as follows:

$$\mathcal{L}_{reg} = \lambda \sum_{i=1}^{N} ||\theta_i - \theta_g||^2 \tag{4.3}$$

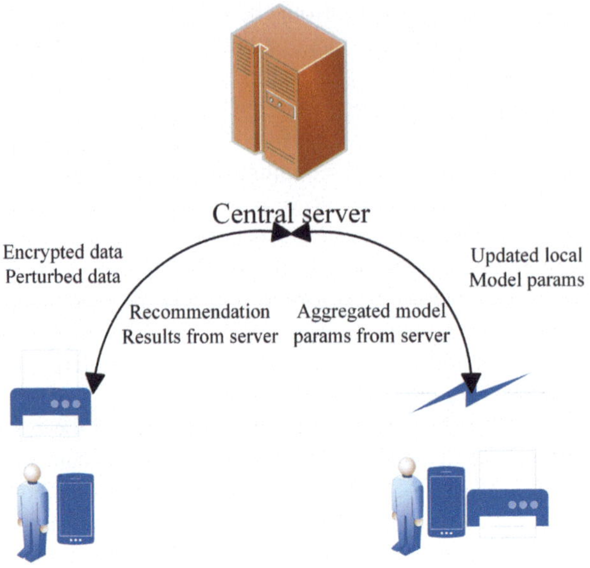

Fig. 4.4 CF-FedSR: adjusts client contributions in federated learning by reweighting aggregation based on local performance and data size aiming to improve model robustness

where λ is a regularization parameter, θ_i represents the local model parameters of client i, θ_g denotes the global model parameters, and N is the number of clients. This term penalizes large deviations between local and global parameters, encouraging a closer alignment and reducing the disparity in updates. By incorporating this regularization term, the federated learning system can maintain a balance between global and local learning objectives, ensuring that all clients benefit from the collaborative training process. This approach is particularly beneficial in scenarios with significant data heterogeneity, where maintaining a balance between global generalization and local specialization is key to achieving fair outcomes.

To ensure comprehensive fairness, it is also important to consider the long-term effects of federated learning updates. Techniques such as fairness-aware curriculum learning can be applied, where clients are gradually introduced into the training process based on their data quality and contribution potential [25, 49]. This phased approach allows the model to progressively adapt to diverse data sources, reducing the risk of bias accumulation over time [29]. By managing the pace and sequence of client participation, fairness-aware curriculum learning ensures that all clients have an equitable opportunity to influence the global model, fostering long-term fairness in federated learning systems [38].

The fairness effects of these methods are usually verified by variance analysis in comparison with baseline models [13]. Such analyses involve measuring the standard deviation, entropy, or other statistical metrics of the clients' performance to ensure that the fairness improvements are significant and consistent [11]. Additionally, fairness metrics can be integrated into the evaluation framework to provide a comprehensive assessment of the model's performance across different client groups [8]. These evaluations help in identifying and addressing any residual biases, ensuring that the implemented fairness strategies effectively reduce disparities among clients.

In conclusion, achieving client-level fairness in federated learning involves addressing both server-side and client-side biases. By reweighting parameter aggregation and incorporating regularization terms, federated learning systems can ensure a more uniform performance across all clients, mitigating the impact of data imbalances and preference divergences. Continued research and innovation in this area will be crucial for developing fair and effective federated learning models that benefit all participants equitably. The exploration of new fairness metrics, adaptive aggregation techniques, and long-term fairness strategies will be essential for advancing the field and ensuring that federated learning systems are both inclusive and robust. As federated learning continues to evolve, ensuring client-level fairness will remain a key challenge and area of focus, driving the development of more equitable and reliable models for a diverse range of applications.

4.2.2 Group-Level Fairness

Group-level fairness aims to mitigate performance bias against protected disadvantaged or minority groups. In federated learning, where client data can be highly heterogeneous, ensuring that the model does not disproportionately favor or disadvantage any group is crucial. The definition of fairness varies depending on how groups are divided and which fairness notions are desired. Groups may be based on demographic attributes, system capabilities, or other relevant characteristics. The objective is to ensure equitable performance across all groups, regardless of differences in data or resources.

F2MF [9] introduces a fairness approach that minimizes discrepancies in group-average metrics between group pairs. This method addresses group-level bias by ensuring that performance metrics for different groups are as close as possible. The fairness loss is jointly optimized with the target recommendation task as part of a multi-task learning framework, balancing accuracy and fairness. To further protect sensitive information, differential privacy is used to obfuscate group affiliations, ensuring that they do not influence the learning process.

Similarly, FPFR [50] defines a fairness loss between two client groups, C_1 and C_2, as follows:

$$L_{fair} = \left| \frac{1}{|C_1|} \sum_{c \in C_1} \mathrm{Eval}(f_c) - \frac{1}{|C_2|} \sum_{c' \in C_2} \mathrm{Eval}(f_{c'}) \right| \tag{4.4}$$

where f_c represents the local model of client c and NDCG@10 is used as the evaluation metric. This loss measures the performance difference between groups, ensuring that the model provides fair recommendations across groups. FPFR uses k-means clustering on user embeddings to group clients with similar characteristics, facilitating an accurate assessment of fairness.

In contrast, RF^2 [35] categorizes clients into low, mid, and high tiers based on system capabilities, addressing performance variability due to resource disparities. This tiered approach allows for federated fairness to be evaluated across different system capabilities, ensuring that the model adapts to the needs of clients with varying resources.

F2MF [9], through differential privacy, ensures that group affiliations are hidden, maintaining both fairness and data confidentiality. The multi-task learning framework used in F2MF optimizes both fairness and recommendation accuracy simultaneously, ensuring that no group is disproportionately advantaged or disadvantaged.

FPFR [50] explicitly measures fairness loss across client groups and uses k-means clustering to improve the accuracy of fairness assessments. The use of NDCG@10 as the evaluation metric strengthens the model's ability to offer personalized, fair recommendations to different user groups.

Finally, RF^2 [35] addresses system-induced heterogeneity by grouping clients into tiers based on computational capabilities. This stratified approach ensures that clients with varying levels of resources are treated fairly and contribute equitably to the global model.

Group-level fairness can also involve techniques that ensure transparency and accountability, such as fairness-aware model auditing. Regular evaluation of model performance across groups can identify and mitigate biases, ensuring that no group is systematically disadvantaged. Additionally, fairness constraints can be incorporated into federated learning algorithms to adjust learning rates or gradient updates, balancing accuracy and fairness across different groups.

Engaging stakeholders in the fairness evaluation process is another critical aspect. Input from different groups can help to design more inclusive federated learning systems, ensuring that the model aligns with the needs of diverse user communities. By integrating fairness checks and involving stakeholders, federated learning systems can be made more socially responsible and equitable.

In summary, group-level fairness in federated learning requires addressing biases that affect different demographic or system-defined groups. Techniques like differential privacy, multi-task learning, and tiered stratification based on system capabilities ensure equitable performance across groups. Continued research and innovation are essential for developing fair and effective federated learning systems.

4.2.3 Summary and Challenges

It is expected that the empowerment of fairness perception will not cause adverse effects on the overall performance of federated models. The empirical study in F2MF [9] indicates a trade-off between fairness and accuracy, highlighting the complexity of achieving both objectives simultaneously. Implementing fairness-aware mechanisms may sometimes lead to a slight decrease in model accuracy, which must be carefully managed to maintain a high level of performance. This trade-off underscores the need for careful calibration of fairness constraints to ensure that they do not overly penalize the model's predictive capabilities, thereby diminishing its utility.

Moreover, the evolutionary nature of users' interests presents a significant challenge. Groups divided by static attributes cannot completely represent disadvantaged or minority parties over time. Static grouping may fail to capture the dynamic changes in user preferences, leading to outdated or inaccurate fairness assessments. Therefore, federated learning systems need to incorporate mechanisms that can dynamically adjust group definitions based on evolving user data. This dynamic adjustment could involve periodic reassessment of group boundaries or the use of machine learning models that continuously learn and adapt to changes in user behavior, ensuring that fairness assessments remain relevant and effective.

Simultaneously, the dynamic data heterogeneity induced by client sampling and non-IIDness (non-Independent and Identically Distributed data) also limits local debiasing efforts on the fair performance distribution [10]. Non-IID data can cause significant variations in local models, making it difficult to achieve uniform performance across all clients. This variability necessitates robust algorithms that can handle data heterogeneity while ensuring fair treatment of all clients. Developing such algorithms requires a deep understanding of the underlying data distributions and the potential biases they introduce, as well as the ability to design models that can adapt to these challenges without sacrificing fairness.

Addressing client- and server-side biases requires comprehensive strategies. On the client side, developing adaptive algorithms that can respond to changing user behaviors and data distributions is crucial. These algorithms should be capable of real-time adjustments to maintain fairness without compromising accuracy. For instance, reinforcement learning techniques could be employed to continuously adjust model parameters based on feedback from user interactions, ensuring that the system remains responsive to changes in the data environment. On the server side, implementing aggregation techniques that consider both the size and quality of local datasets can help mitigate biases. Aggregation strategies could be enhanced by incorporating fairness-aware weighting schemes that give more influence to underrepresented or disadvantaged groups, thus ensuring a more balanced representation in the global model. Additionally, employing fairness-aware federated learning frameworks that balance the contributions of all clients, regardless of their data volume or computational power, is essential. These frameworks should be designed to promote equity among clients, ensuring that the federated learning process does not disproportionately benefit or harm any particular group.

Furthermore, incorporating fairness into federated learning involves balancing multiple fairness objectives, such as demographic parity, equal opportunity, and equalized odds. Each of these objectives requires different techniques and considerations, making it challenging to develop a one-size-fits-all solution. Researchers need to explore multi-objective optimization frameworks that can simultaneously address various fairness criteria. Such frameworks would allow for the simultaneous optimization of multiple fairness metrics, enabling more comprehensive and nuanced fairness assessments. This approach could involve the use of Pareto optimization techniques, where trade-offs between different fairness objectives are carefully managed to achieve an optimal balance that meets the overall fairness goals of the system.

In addition, exploring the intersection of fairness with other critical concerns in federated learning, such as privacy and security, presents another layer of complexity. For example, achieving fairness while also maintaining strong privacy guarantees (e.g., through differential privacy) can be particularly challenging, as the addition of noise to preserve privacy may exacerbate biases or reduce the effectiveness of fairness interventions. Therefore, research into privacy-preserving fairness algorithms is essential, ensuring that the goals of fairness and privacy are pursued in tandem rather than at odds with each other.

Another important area of exploration is the development of fairness-aware federated learning systems that can operate effectively in the presence of adversarial attacks. Adversarial participants may attempt to manipulate the federated learning process to undermine fairness, for example, by injecting biased data or disrupting the aggregation process. Developing robust defenses against such attacks is crucial to maintaining the integrity and fairness of the system.

In conclusion, achieving fairness in federated learning is a multifaceted challenge that involves addressing both static and dynamic biases, managing trade-offs between fairness and accuracy, and developing adaptive and robust algorithms. As federated learning continues to evolve, ongoing research and innovation will be critical in creating systems that are both fair and effective, ensuring that all users benefit equitably from the advances in machine learning technology. The pursuit of fairness must be holistic, considering the interplay between fairness, privacy, security, and performance to create federated learning systems that are not only technically sound but also socially responsible. Addressing these challenges will require collaboration across disciplines, drawing on insights from machine learning, ethics, law, and social sciences to develop comprehensive solutions that meet the diverse needs of all stakeholders in federated learning environments.

References

1. Abebe R, Barocas S, Kleinberg J, Levy K, Raghavan M, Robinson D (2019) Fairness and abstraction in sociotechnical systems. In: Proceedings of the 2019 ACM conference on fairness, accountability, and transparency (FAT)
2. Al-Fuqaha A, Guizani M, Mohammadi M, Aledhari M, Ayyash M (2015) A survey on internet of things: architecture, enabling technologies, security and privacy, and applications. IEEE Commun Surv Tutorials 17(4):2347–2376
3. Bell RM, Koren Y, Volinsky C (2002) The effect of sparsity on collaborative filtering. In: Proceedings of the 8th ACM SIGKDD international conference on knowledge discovery and data mining (KDD)

4. Chai Y, Wu F, Zhang H, Zhang J, Zhou Z (2020) FedIS: federated incomplete service recommendation. IEEE Trans Serv Comput 13(5):1010–1023
5. Chai Y, Xiang Y, Yuan W, Chen G, Zhang Q (2020) LightFR: lightweight federated recommendation with efficient privacy-preserving matrix factorization. In: Proceedings of the 14th ACM conference on recommender systems (RecSys)
6. Chen C, Chen M, Zeng M, Wang Y, Deng J (2020) FedMeta: a meta-learning framework for federated learning. In: Proceedings of the 29th ACM international conference on information and knowledge management (CIKM)
7. Cheng X, Zhao M, Qi J, Wang C, Li X (2020) Hierarchical federated semi-asynchronous aggregation for personalized mobile recommendation. In: Proceedings of the 29th ACM international conference on information and knowledge management (CIKM)
8. Donahue K, Kleinberg J (2020) Fairness and abstraction in sociotechnical systems. In: Proceedings of the ACM conference on fairness, accountability, and transparency
9. Du Y, Zhang J, Liu J, Wang S (2020) F2MF: fairness-aware federated matrix factorization for recommendations. In: Proceedings of the 29th ACM international conference on information and knowledge management (CIKM)
10. Du Y, Zhang J, Liu J, Wang S (2021) FairFed: enabling group fairness in federated learning. In: Proceedings of the 29th ACM international conference on information and knowledge management (CIKM)
11. Dwork C, Roth A (2012) The algorithmic foundations of differential privacy. Found Trends Theor Comput Sci 9(3–4):211–407
12. Gou J, Yu B, Maybank SJ, Tao D (2021) A comprehensive survey on knowledge distillation: from response to feature and relation. IEEE Trans Neural Netw Learn Syst 32(12):4816–4832
13. Hard A, Rao K, Mathews R, Beaufays F, Augenstein S, Eichner H, Kiddon C (2018) Federated learning for mobile keyboard prediction. Preprint. arXiv:1811.03604
14. Hidasi B, Karatzoglou A, Baltrunas L, Tikk D (2015) Session-based recommendations with recurrent neural networks. In: Proceedings of the 27th ACM international conference on information and knowledge management (CIKM)
15. Hinton G, Vinyals O, Dean J (2015) Distilling the knowledge in a neural network. In: Proceedings of the 28th international conference on neural information processing systems (NeurIPS)
16. Huang J, Wang W, Zhao L, Liu X, Xu B (2020) PreffedPOI: preference-aware federated learning for personalized POI recommendation. In: Proceedings of the 28th ACM international conference on information and knowledge management (CIKM)
17. Kairouz P, McMahan HB, Alistarh D, Avestimehr S, Balcan MF et al (2019) Towards federated learning at scale: system design. Preprint. arXiv:1902.01046
18. Li Y, Guo Y, Jiang Y, Chen L, Liu X (2019) Federated collaborative autoencoder model selection for recommendation. In: Proceedings of the 28th ACM international conference on information and knowledge management (CIKM)
19. Li T, Sahu AK, Sanjabi M, Talwalkar A (2020) Fairness in federated learning: challenges and opportunities. In: Proceedings of the 29th ACM international conference on information and knowledge management (CIKM)
20. Li T, Sahu AK, Zaheer M, Sanjabi M, Talwalkar A (2020) FedFast: efficient federated learning via centralized and decentralized group communication. In: Proceedings of the 5th international conference on learning representations (ICLR)
21. Lin Y, Han S, Mao H, Wang Y, Dally WJ (2020) Federated knowledge distillation for on-device learning. In: Proceedings of the 5th international conference on learning representations (ICLR)
22. Liu H, Wang J, Shen S, Wang X (2019) ReFRS: region-aware federated recommendation system for points of interest. In: Proceedings of the 28th ACM international conference on information and knowledge management (CIKM)
23. Liu H, Wu Q, Ding G, He Z, Lin Z (2019) Unsupervised recurrent federated learning for spatio-temporal data. In: Proceedings of the 28th ACM international conference on information and knowledge management (CIKM)

24. Liu X, Wang T, Chen J, Tang M, Xie M, Xu W (2020) A survey on federated learning in internet of things: challenges, solutions, and future directions. IEEE Internet Things J 7(12):11293–11313
25. Liu Y, Wu X, Liu W, Li Z (2021) FairFed: fair federated learning for unbiased data aggregation. In: Proceedings of the AAAI Conference on Artificial Intelligence
26. Lu C, Xu G, Li Z, Zhang Y, Lin X (2019) Cali3F: client allocation for fair federated learning in financial applications. In: Proceedings of the 28th ACM international conference on information and knowledge management (CIKM)
27. Luo S, Zheng L, Wang J, Jiang Y, Xu S (2020) Hierarchical federated collaborative filtering with compression and differential privacy. In: Proceedings of the 26th ACM SIGKDD conference on knowledge discovery and data mining (KDD)
28. Mehrabi N, Morstatter F, Saxena N, Lerman K, Galstyan A (2021) A survey on bias and fairness in machine learning. ACM Comput Surv 54(6):115
29. Mohri M, Sivek G, Suresh AT (2019) Agnostic federated learning. Preprint. arXiv:1902.00146
30. Nishio T, Yonetani R (2020) Payload optimization method for federated learning. In: Proceedings of the 29th ACM international conference on information and knowledge management (CIKM)
31. Seide F, Fu H, Droppo J, Li G, Yu D (2014) One-bit stochastic gradient descent and its application to data-parallel distributed training of speech DNNs. In: Proceedings of the 15th annual conference of the international speech communication association (INTERSPEECH)
32. Shi W, Cao J, Zhang Q, Li Y, Xu L (2016) Edge computing: vision and challenges. IEEE Internet Things J 3(5):637–646
33. Tang X, Chen G, Huang C, Zhang P, Gu X (2021) PerFedRec++: enhancing personalized federated recommendation with adaptive user embeddings. In: Proceedings of the 29th ACM international conference on information and knowledge management (CIKM)
34. Tang X, Chen G, Huang C, Zhang P, Gu X (2021) PerFedRec: personalized federated recommendation with visual explanation. IEEE Trans Knowl Data Eng 33(11):4001–4014
35. Tang X, Chen G, Huang C, Zhang P, Gu X (2021) RF2: robust fair federated learning with tiered groups. In: Proceedings of the 29th ACM international conference on information and knowledge management (CIKM)
36. Wang J, Zhang T, Sebe N, Shen HT (2016) Learning to hash for indexing big data: a survey. Proc IEEE 104(1):34–57
37. Wang X, Sun Y, Wang J, Liu J (2019) STD-CN: a novel learning paradigm for efficient edge computing with spatio-temporal data analysis. In: Proceedings of the 28th ACM international conference on information and knowledge management (CIKM)
38. Wang J, Liu Q, Liang H, Joshi G, Poor HV (2020) Tackling the objective inconsistency problem in heterogeneous federated optimization. In: Advances in neural information processing systems
39. Wang M, Zhang H, Fu H, Cao L (2020) JointRec: A joint cloud-edge approach for communication-efficient federated recommendation. In: Proceedings of the 43rd international ACM SIGIR conference on research and development in information retrieval (SIGIR)
40. Wu C, Sun F, Guo Z, Zhang Q, Liu L (2020) SemiDFEGL: semi-supervised federated learning with differentially private edge-graph learning. In: Proceedings of the 29th ACM international conference on information and knowledge management (CIKM)
41. Xia R, Wu Y, Zhang J (2020) Edge-cloud collaborative training for personalized federated learning. IEEE Trans Parallel Distrib Syst 31(6):1294–1306
42. Xie Y, Liu Z, Tang X, Zhang Q, Liu C (2020) GIMA: a game-theoretic incentive mechanism for mobile edge computing and federated learning. IEEE Internet Things J 7(7):6402–6413
43. Yi J, Wu F, Wu C, Liu R, Sun G, Xie X (2021) Efficient-FedRec: efficient federated learning framework for privacy-preserving news recommendation, pp 2814–2824. https://doi.org/10.18653/v1/2021.emnlp-main.223, https://aclanthology.org/2021.emnlp-main.223
44. Yu X, Li J, Zhang Y, Zhou Y, Wang G (2021) CF-FedSR: collaborative filtering-based federated sequential recommendation. IEEE Trans Knowl Data Eng 33(11):4001–4014

45. Yue M, Zhang W, Zhang Y, Feng J (2022) A survey on fair federated learning. IEEE Trans Big Data 8(1):1–21
46. Zhang H, Zhang J, Huang HV, Liu L, Shen XS (2015) A two-stage stackelberg game for joint resource allocation and pricing in cloud computing networks. IEEE Trans Netw Serv Manage 12(3):400–413
47. Zhang Y, Wang H, Liu Q, Li T (2019) CCF-CRR: context-aware collaborative filtering with contextualized route recommendation. In: Proceedings of the 28th ACM international conference on information and knowledge management (CIKM)
48. Zhang J, Liu J, Wang S, Wang X (2021) CPF-POI: cluster-based personalized federated learning for POI recommendation. IEEE Trans Knowl Data Eng 33(11):4001–4014
49. Zhang J, Wu C, Wu F (2021) FairFL: fair federated learning via reinforcement learning-based client scheduling. In: Proceedings of the AAAI conference on artificial intelligence
50. Zhao L, Chen X, Li M, Xu B (2021) FPFR: fair personalized federated recommendation with group-level fairness. IEEE Trans Knowl Data Eng 33(11):4001–4014
51. Zhao L, Liu X, Zhao Z, Li M, Xu B (2021) CdFed: collaborative federated learning for cross-domain recommendation. IEEE Trans Knowl Data Eng 33(11):501–526
52. Zhou Z, Cao Y, Dong H, Liu C (2018) A survey of mobile edge computing in the industrial internet of things: requirements, architecture, and key technologies. IEEE Internet Things J 5(2):1205–1222
53. Zhou Z, Feng S, Zhang K, Yang Q (2020) Federated edge learning with resource-constrained devices: optimization, communication reduction, and privacy. IEEE Internet Things J 7(12):10240–10253
54. Zhu Q, Zhou Y, Liu G, Xue M (2020) FL-GMT: federated learning with global model trust for non-IID data. In: Proceedings of the 29th ACM international conference on information and knowledge management (CIKM)

Chapter 5
Future Prospects

Abstract This chapter explores the future of cross-device federated recommendation systems, focusing on benchmarks, heterogeneous federated learning, knowledge graphs, and encryption techniques. It highlights the need for standardized protocols for fair evaluations and addresses model adaptability, communication efficiency, and equity in heterogeneous settings. The chapter discusses knowledge graphs for improved node representations and advanced encryption for enhanced data security. It provides insights and outlines research directions to ensure scalability and applicability in real-world scenarios.

5.1 Specialized Cross-Device Federated Benchmark

In centralized recommendation systems, standardized experimental settings and performance comparisons have been systematically established through widely recognized benchmarks such as RecBole [43, 44] and FuxiCTR [47, 48]. These benchmarks provide a common framework for evaluating models, ensuring that comparisons are conducted under consistent conditions with uniform datasets, evaluation metrics, and experimental protocols. Current open-source projects for HFedRec exhibit significant heterogeneity in programming languages, machine learning libraries, and federated environments. This diversity complicates the direct comparison of approaches and impedes the development of a unified framework for evaluating federated recommendation models. Although several baseline recommendation models (e.g., MF [10, 25, 27], NCF [9], and LightGCN [8]) are widely adopted in the field, they lack a unified federated setting that standardizes critical aspects such as non-IID data distribution, client sampling strategies, client size variability, and global aggregation methods. These elements are essential for a rigorous and comprehensive evaluation of models in federated environments. Given the seamless integration of various recommendation models with federated training procedures, as discussed in Chap. 2, there is an immediate and significant need to establish a benchmark specifically designed for HFedRec (or FedRec). Such a benchmark would serve as a foundational tool for advancing the field, enabling fair

X. Kong et al., *Cross-device Federated Recommendation*, Machine Learning: Foundations, Methodologies, and Applications,
https://doi.org/10.1007/978-981-96-3212-1_5

and transparent comparisons of models, and providing a structured framework for assessing communication costs, fairness, and other essential aspects of federated learning. To address this critical need, it is imperative to develop a specialized benchmark for cross-device federated recommendation systems. This benchmark should include several key components to ensure comprehensive evaluation and comparability across different models and settings:

- **Standardized Data Distribution and Partitioning:** A fundamental component of the proposed benchmark is the establishment of standardized protocols for data partitioning and distribution. In federated settings, data is inherently non-IID, reflecting the diverse and decentralized nature of real-world environments. The benchmark must define precise protocols for handling non-IID data, ensuring that data partitioning among clients realistically simulates the challenges encountered in practical applications. These protocols should provide detailed guidelines on how to manage and distribute data among clients, fostering a consistent and reproducible approach across different studies.
- **Uniform Federated Learning Protocols:** To facilitate fair and meaningful comparisons across models, it is essential to establish standardized federated learning protocols. These protocols should cover various aspects, including client sampling strategies, client size variability, and global aggregation methods. Adherence to these standardized protocols ensures that models are evaluated under consistent conditions, thereby eliminating variability arising from differing experimental setups. This uniformity is crucial for drawing reliable and generalizable conclusions from comparative studies.
- **Comprehensive Performance Metrics:** A holistic evaluation of federated recommendation models requires a broad spectrum of performance metrics beyond accuracy. The benchmark should include metrics that assess communication costs, computational efficiency, fairness, and scalability. These additional metrics are particularly relevant in federated settings, where communication overhead and the equitable treatment of clients are critical considerations. By incorporating these metrics, the benchmark provides a comprehensive view of model performance, enabling researchers to evaluate trade-offs and optimize their models accordingly.
- **Extensible and Reproducible Framework:** For the benchmark to remain relevant and useful over time, it must be designed with extensibility and reproducibility as core principles. The benchmark framework should be sufficiently flexible to accommodate new models, methods, and technologies as they emerge. Additionally, the framework should include clear and detailed documentation, along with well-structured code repositories, to ensure the reproducibility of experiments. Emphasizing reproducibility allows other researchers to validate findings, build on existing work, and contribute to the collective advancement of the field.
- **Diverse Benchmarking Scenarios and Datasets:** To ensure the benchmark's applicability across a wide range of federated recommendation problems, it should incorporate a variety of benchmarking scenarios and datasets. These

scenarios should reflect different application domains, user behaviors, and data characteristics, capturing the diverse challenges that arise in real-world federated environments. By providing a rich and diverse set of benchmarking scenarios, the benchmark enables the development of models that are robust, generalizable, and capable of addressing the specific needs of various application contexts.
- **Collaborative Development and Maintenance:** Finally, the benchmark should be developed and maintained through a collaborative effort within the research community. This collaborative approach ensures that the benchmark evolves in response to emerging challenges, incorporates the latest advancements in federated learning and recommendation systems, and remains aligned with the research community's needs and priorities. By fostering a culture of collaboration, the benchmark becomes a dynamic resource that continues to drive progress in the field.

By establishing a specialized cross-device federated benchmark, the research community can systematically address the unique challenges posed by federated recommendation systems. This benchmark not only facilitates fair and rigorous comparisons of models but also provides a structured framework for advancing the state of the art. In doing so, it lays the foundation for developing more effective, efficient, and equitable federated recommendation systems, driving significant progress in this rapidly evolving field.

5.2 Exploration of Heterogeneous Federated Learning

Most existing HFedRec frameworks implicitly assume homogeneous local recommendation model architectures across all participants, which significantly restricts their applicability in real-world scenarios. In practice, this assumption often proves unrealistic. For example, resource-constrained clients, such as mobile devices or edge nodes, may lack the computational power to support sophisticated, resource-intensive models. Consequently, these clients must be excluded from the federated learning environment if the collaboratively trained model becomes too complex, thereby undermining the inclusivity and scalability that federated learning aims to achieve.

To address this issue, heterogeneous federated learning (HFL) has emerged as a promising solution [34]. HFL allows different clients to maintain models that vary in size and architecture, a concept referred to as model heterogeneity. This approach enables a more inclusive federated learning ecosystem where clients with varying computational capabilities can participate. Recent research has delved into the transmission of various types of knowledge carriers, such as logits on public datasets [15], additional homogeneous models [24, 28], and class prototypes [30], to facilitate knowledge sharing across locally heterogeneous models. These strategies enable different clients to benefit from shared knowledge without requiring uniform model structures.

However, despite the potential of these methodologies, they cannot be directly applied to horizontal federated recommendation systems due to the distinct data structures, model architectures, and parameters involved. The unique characteristics of recommendation systems, such as the need to manage large and sparse user-item interaction matrices, pose additional challenges that are not present in other federated learning domains. For instance, traditional HFL approaches may not adequately address the complexity of integrating diverse recommendation models that differ not only in architecture but also in the way they process and represent user and item data.

Yuan et al. [37] represents a pioneering effort in model-heterogeneous HFedRec, allowing each client to maintain item embeddings of varying sizes. This flexibility is a significant step forward, as it enables clients with different resource constraints to participate in the federated learning process. However, HeteFedRec still requires that each client employs the same local model architecture, which limits its ability to fully realize the potential of heterogeneous federated learning. The reliance on a uniform model architecture means that clients with limited resources may still be unable to participate fully, as they might not be able to support the required model complexity.

To develop truly heterogeneous federated learning frameworks, several critical challenges must be addressed. First and foremost, effective aggregation strategies are essential to handle the diversity in model architectures and sizes. Traditional federated learning approaches rely on aggregating model parameters or gradients, but this becomes significantly more complex when the models differ in structure. Aggregating knowledge from heterogeneous models without a common architecture demands novel techniques that can align and integrate disparate model parameters and outputs. This may involve advanced methods such as meta-learning or neural architecture search (NAS) [50] to dynamically adapt models for effective collaboration. Meta-learning, for instance, can help in creating a universal model that can generalize across different client-specific models, while NAS can be used to optimize model architectures based on the specific constraints and capabilities of each client.

Communication efficiency is another critical concern in heterogeneous federated learning. The heterogeneity of client models exacerbates the challenges of communication, as model updates and knowledge carriers vary in size and complexity. Optimizing the communication of these updates is crucial to minimizing bandwidth and latency issues, which are more pronounced in a heterogeneous environment. Techniques such as gradient compression, quantization, and efficient encoding of knowledge carriers can play a pivotal role in mitigating these challenges. For example, gradient compression reduces the amount of data transmitted by approximating the gradients, while quantization reduces the precision of the model parameters, thereby lowering the communication overhead. Efficient encoding methods can also be employed to ensure that the transmitted knowledge retains its utility while minimizing the data size.

Ensuring fairness and robustness in federated learning scenarios with heterogeneous models is also paramount. The variability in client capabilities and model architectures introduces the risk of biased learning outcomes, where more powerful clients may disproportionately influence the global model. Mechanisms must be developed to ensure equitable contribution and benefit for all clients, regardless of their resource constraints, to ensure the ethical deployment of federated learning systems. This could involve techniques such as weighted aggregation, where the contributions of clients are balanced based on their capabilities, or adversarial training methods that enhance the robustness of the models against potential attacks.

Moreover, real-world applications of heterogeneous federated learning must consider the privacy and security implications of model heterogeneity. Ensuring that the transmission of knowledge carriers and model updates does not compromise client data privacy is crucial. In a heterogeneous setting, where different clients may have varying levels of privacy requirements and computational resources, maintaining data privacy becomes even more challenging. Advanced privacy-preserving techniques such as differential privacy and secure multi-party computation should be integrated into HFL frameworks to uphold stringent privacy standards. Differential privacy, for instance, ensures that the information shared during the learning process does not reveal specific details about any individual client's data, while secure multi-party computation allows multiple clients to jointly compute a function over their inputs without revealing their private data to each other.

In conclusion, the exploration of heterogeneous federated learning represents a critical frontier in the evolution of federated recommendation systems. By embracing the diversity of client capabilities and model architectures, researchers can develop more inclusive and effective federated learning frameworks that reflect the complexity of real-world environments. As the field continues to advance, the insights gained from model-heterogeneous HFedRec will undoubtedly contribute to the broader understanding of federated learning and its applications, paving the way for more robust, scalable, and privacy-preserving systems that benefit a wide range of users and stakeholders.

5.3 Beyond Plain Graph with Knowledge

Incorporating side information about users or items is a well-established approach to enhancing recommendation performance, and knowledge graphs (KGs) are particularly effective in this regard [6]. A knowledge graph is a structured data representation consisting of nodes and edges, where nodes symbolize entities (such as users or items) or abstract concepts, and edges represent the relationships between these entities or concepts. Each edge in a KG is typically described by a triplet (*head*, *relation*, *tail*), which articulates the relationship from the head entity to the tail entity through the specified relation. By capturing and expressing intricate relationships, knowledge graphs enrich node representations, thereby providing more nuanced and insightful contextual information that can significantly boost the performance of recommendation systems.

The ability of knowledge graphs to model complex relationships within a dataset is crucial, particularly in domains where user-item interactions are sparse or where traditional collaborative filtering methods struggle to capture the full breadth of contextual information. For example, in a movie recommendation system, a KG could connect a user not only to the movies they have watched but also to the actors, directors, and genres associated with those movies, thereby enabling the system to make recommendations based on a more comprehensive understanding of user preferences. This approach can mitigate the limitations of sparse interaction data by supplementing it with rich relational data, ultimately leading to more accurate and personalized recommendations.

Despite their advantages, the construction and maintenance of KGs pose significant challenges, particularly in terms of data privacy and integration across different organizations. Different KGs within the same domain can complement each other, providing a more holistic view of the data. However, due to concerns over data privacy and the competitive nature of business interests, transferring triplets or other sensitive data to a centralized repository is often infeasible or outright prohibited. This constraint limits the potential of KGs to fully exploit the collective knowledge across distributed datasets.

The advent of federated knowledge graph (FedKG) learning offers a promising solution to these challenges, enabling collaborative representation learning of KGs without compromising the privacy of the underlying data. FedKG extends the principles of federated learning to the domain of knowledge graphs, allowing multiple organizations to collaboratively train models on their combined knowledge graphs without sharing the actual data. This approach preserves the privacy of individual datasets while still enabling the benefits of shared learning. For instance, Chen et al. [4] have made significant strides in this field by enhancing global entity embeddings using a parameter server (PS) architecture. This method facilitates the sharing of entity embeddings across different clients while keeping the underlying knowledge graphs decentralized. Similarly, Peng et al. [23] have implemented federated KG embedding through the PATE-GAN model [13] in a decentralized manner, ensuring that knowledge is shared without direct data exchange. These advancements in FedKG represent a significant leap forward, enabling enhanced knowledge sharing and more robust recommendation models without risking data leakage.

The integration of HFedRec and FedKG holds substantial potential for further advancing the performance and applicability of federated recommendation systems. By leveraging the complementary strengths of both federated learning and knowledge graphs, it is possible to achieve superior recommendation outcomes that are both more accurate and more personalized. Specifically, FedKG can provide richer contextual information about users and items, which can be used to inform the federated learning process, leading to more nuanced and tailored recommendations. For example, in a federated learning environment where data is distributed across multiple clients, the integration of a KG can help bridge the gaps between isolated data silos by providing a common framework for understanding the relationships between users and items.

Moreover, the fusion of HFedRec and FedKG can address some of the inherent limitations of traditional federated recommendation systems. One of the primary challenges in federated recommendation is the sparsity of user-item interactions, which can limit the ability of the model to make accurate predictions. The incorporation of KG data can help mitigate this issue by providing additional relational data that enhances the model's understanding of user preferences and item characteristics. This enriched representation allows the recommendation system to make more informed predictions, even in the presence of sparse interaction data. For instance, if a user has interacted with only a few items, the KG can infer additional preferences based on the relationships between those items and other entities in the graph, leading to more robust and reliable recommendations.

Furthermore, the development of FedKG technology is likely to stimulate new research directions within federated learning. Researchers can explore novel methods for efficient and secure knowledge sharing across distributed networks, addressing challenges such as data heterogeneity, scalability, and communication efficiency. The heterogeneous nature of the data used in KGs, which often spans different domains and formats, presents unique challenges for federated learning. Advanced techniques such as differential privacy, secure multi-party computation, and federated transfer learning can be applied to enhance both the privacy and performance of FedKG systems. For example, differential privacy can be used to ensure that the shared embeddings or model parameters do not inadvertently reveal sensitive information about the underlying data. Secure multi-party computation can enable multiple organizations to jointly compute shared models or embeddings without exposing their individual data, while federated transfer learning can facilitate the application of pre-trained models to new domains or datasets, further enhancing the flexibility and applicability of FedKG systems.

In addition to these technical advancements, the integration of HFedRec and FedKG could lead to significant improvements in the user experience. By leveraging the rich relational data provided by KGs, recommendation systems can offer more relevant and personalized suggestions, which are likely to resonate more strongly with users. This could result in higher user satisfaction and engagement, as well as better retention rates for services that rely on recommendation systems. Moreover, the privacy-preserving nature of FedKG ensures that these benefits can be realized without compromising user trust or violating data privacy regulations.

In summary, the integration of HFedRec and FedKG represents a promising avenue for advancing the field of federated recommendation systems. By harnessing the power of knowledge graphs, these systems can deliver more accurate, personalized, and privacy-preserving recommendations, paving the way for innovative applications and improved user experiences. The ongoing research and development in this area are likely to yield significant benefits for both academia and industry, driving the evolution of next-generation recommendation technologies. As the field continues to mature, the insights gained from the integration of federated learning and knowledge graphs will play a crucial role in shaping the future of recommendation systems, making them more effective, inclusive, and secure.

5.4 Efficient Encryption Techniques

Secure multi-party computation (SMC) consumes significant data exchanges among participants, especially when multiple privacy protocols are involved in cooperation [1, 3]. This substantial data exchange can lead to increased communication overhead and latency in federated learning systems. On the other hand, differential privacy (DP), while effective in certain scenarios, shows vulnerabilities when faced with sophisticated malicious attacks [26, 38]. As a result, homomorphic encryption (HE) has emerged as a preferred choice for ensuring data privacy in federated learning due to its capability to perform computations on encrypted data without decryption. HE's ability to preserve the privacy of sensitive data while still allowing meaningful computations to be performed on it makes it a powerful tool in the context of federated learning.

Despite its advantages, HE poses challenges in practical implementations, particularly concerning the significant inflation of plaintext size. According to Jiang et al. [12], the Paillier encryption scheme [22] inflates plaintext size by approximately 0.2k×, while the CKKS scheme [5] results in a staggering 173k× increase. This considerable expansion in data size can lead to memory overflow risks and impose heavy computational burdens on resource-constrained clients, thereby limiting the scalability and efficiency of HE in federated learning. The inflated data size not only increases the storage requirements but also slows down the encryption and decryption processes, making it difficult to deploy HE in large-scale systems.

To address these challenges, BatchCrypt [39] has been proposed as a solution that enables batch encryption. This technique allows the concurrent encryption of multiple plaintexts, thereby reducing the overall length of the resulting ciphertexts. By optimizing the encryption process, BatchCrypt enhances the efficiency of HE, making it more feasible for use in large-scale federated learning scenarios. BatchCrypt achieves this by packing multiple messages into a single ciphertext, which significantly reduces the communication overhead and improves the computational efficiency of the encryption process. This method ensures that the benefits of HE can be leveraged without overwhelming the system's resources. The ability to perform batch encryption is particularly valuable in federated learning, where large amounts of data need to be encrypted and transmitted frequently.

In addition to BatchCrypt, Liu et al. [12] have introduced a novel homomorphic encryption scheme called FLASHE. This scheme is specifically designed to achieve faster encryption and decryption speeds while meeting the minimum security and functionality requirements for federated learning. FLASHE focuses on secure aggregation at the server side, which is a critical aspect of federated learning where encrypted data from multiple clients are aggregated to update the global model. FLASHE employs a lightweight encryption mechanism that reduces the computational complexity and memory requirements, making it suitable for deployment in real-world federated learning environments. By optimizing the trade-off between encryption strength and computational overhead, FLASHE provides a practical solution for secure and efficient data aggregation. The development of FLASHE highlights the ongoing efforts to make HE more practical for use in federated learning by addressing the key challenges of computational efficiency and data size.

Both BatchCrypt and FLASHE are implemented in the Federated AI Technology Enabler (FATE) framework,[1] demonstrating their practical applicability in real-world federated learning systems. The integration of these techniques within the FATE framework underscores their potential to enhance the scalability and security of federated learning applications. These advancements in homomorphic encryption techniques present a promising direction for enhancing the efficiency and security of federated recommendation systems. By providing robust encryption solutions that do not compromise on performance, these methods enable the practical deployment of secure federated learning systems at scale. The FATE framework serves as a valuable platform for testing and validating new encryption techniques, ensuring that they are both effective and practical for real-world applications.

Moreover, recent developments in lattice-based cryptography have further improved the feasibility of HE for federated learning. Schemes like BFV [7] and CKKS [5] offer support for approximate arithmetic, which is particularly useful for machine learning tasks that involve floating-point computations. These lattice-based schemes provide a strong security foundation while enabling efficient homomorphic operations on encrypted data. By leveraging these cryptographic advancements, federated learning systems can achieve higher levels of security without incurring prohibitive computational costs. The ability to perform approximate arithmetic on encrypted data is particularly important for machine learning, where floating-point operations are common.

In addition to technical improvements, collaborative efforts in the research community are focused on creating standardized protocols and benchmarks for evaluating the performance and security of HE schemes in federated learning. These initiatives aim to provide a clear framework for assessing the trade-offs involved in different encryption methods and to guide the development of more effective privacy-preserving techniques. By establishing common evaluation criteria, researchers and practitioners can better understand the strengths and limitations of various HE approaches and make informed decisions about their implementation in federated learning systems. Standardized benchmarks and protocols are crucial for ensuring that new encryption techniques are rigorously tested and compared against existing methods, facilitating the adoption of the most effective solutions.

Another promising area of research is the combination of HE with other privacy-preserving techniques, such as secure multi-party computation (SMC) and differential privacy (DP). Hybrid approaches that integrate the strengths of multiple techniques can provide more comprehensive privacy guarantees while optimizing for efficiency. For example, combining HE with SMC can enable secure computations on encrypted data while minimizing the data exchange required for multiparty protocols. Similarly, incorporating DP mechanisms can add an

[1] https://fate.fedai.org.

additional layer of protection against inference attacks, ensuring that the privacy of individual data points is preserved even in the presence of sophisticated adversaries. The development of hybrid approaches that combine multiple privacy-preserving techniques is likely to be a key area of innovation in the coming years, offering new ways to balance security, efficiency, and scalability.

By balancing the trade-offs between computational efficiency and security, these innovative encryption methods open new avenues for the development and deployment of robust and scalable HFedRec systems. The continuous improvement in encryption techniques will undoubtedly play a crucial role in the future of privacy-preserving federated learning. As research progresses, it is expected that more sophisticated and efficient encryption schemes will be developed, further enhancing the capability of federated learning systems to operate securely and efficiently in diverse and dynamic environments. The ongoing evolution of encryption techniques will be critical for ensuring that federated learning systems can meet the growing demands of privacy, security, and performance.

In summary, while HE offers a promising solution for ensuring data privacy in federated learning, its practical implementation poses significant challenges related to computational overhead and data size inflation. Techniques like BatchCrypt and FLASHE address these challenges by optimizing the encryption process and reducing the computational burden on clients. The integration of these methods within frameworks like FATE demonstrates their practical applicability and potential to enhance the scalability and security of federated learning systems. Ongoing research and collaboration in the field will continue to drive the development of more efficient and robust encryption techniques, ensuring the secure and scalable deployment of federated learning applications in the future. As the field of federated learning continues to evolve, the development and refinement of efficient encryption techniques will remain a key priority for researchers and practitioners alike, ensuring that federated learning can be deployed securely and effectively at scale.

5.5 Addressing Cold-Start Issue and Federated Scalability

Federated scalability generally refers to the impact of massive participated clients on computing resources, communication efficiency, and multiparty privacy [11, 41]. The scalability challenges become more pronounced as the number of clients increases, necessitating efficient management of computational loads and data exchanges to maintain system performance. As federated learning systems scale up to accommodate a growing number of clients, the demands on both computational and communication resources can become overwhelming. This necessitates the development of more efficient protocols and algorithms that can handle large-scale participation without compromising the overall performance of the system.

In the recommendation scenario, another implication is the cold-start trouble caused by the expanding scale of users and items. The user cold-start (UCS) issue faces new users who have not interacted with enough items to portray their preference, while the item cold-start (ICS) issue copes with how to present new items to users. The cold-start problem is particularly challenging in federated learning environments because the lack of interaction data for new users or items can severely limit the effectiveness of the recommendation model. Without sufficient data to infer preferences, the system struggles to deliver relevant recommendations, leading to a suboptimal user experience.

IFedRec [40] assumes the server holds rich side information of the whole item set. When the ICS issue occurs, clients can utilize encoded item attributes to initialize local item embeddings. This method allows for a more informed initialization process, thereby improving the recommendation quality for new items. By leveraging rich side information, the system can effectively mitigate the challenges posed by the lack of interaction data for new items, ensuring that they are accurately represented in the recommendation model. The use of side information, such as item descriptions, categories, or tags, enables the system to generate more meaningful embeddings for new items, which in turn enhances their visibility and relevance in the recommendation process.

Additionally, Wahab et al. [31] define the ICS problem and design a credibility-based trust mechanism to assess whether clients give reasonable predicted scores to cold items. This trust mechanism helps in filtering out unreliable client predictions, ensuring that the recommendations remain accurate and trustworthy even when dealing with new items. By incorporating a credibility assessment, the system can dynamically adjust the weight given to different clients' predictions, prioritizing more reliable sources and thus enhancing the overall recommendation accuracy. The credibility-based approach is particularly useful in federated learning, where the heterogeneity of client data and behavior can lead to varying levels of prediction accuracy.

Based on the observation that most users have few interactions in real-world news recommendation scenarios, FINDING [36] runs k-means on user representations to cluster clients into groups, thereby benefiting cold users from cluster-level models. By leveraging the cluster-level models, the system can provide better recommendations to new users based on the preferences of similar users within the same cluster. This approach not only addresses the UCS issue but also enhances the overall recommendation accuracy for users with limited interaction data. Clustering similar users together allows the system to infer preferences more accurately, leveraging collective behavior patterns to inform recommendations for new or less active users. This method capitalizes on the notion that users within the same cluster are likely to share similar preferences, making it possible to make more informed predictions even in the absence of extensive individual interaction data.

To improve federated scalability and mitigate cold-start issues, several practical strategies can be implemented. One approach is to integrate more refined machine learning models that specifically target user behavior and item characteristics, which can enhance the quality of initial embeddings and predictions. By using models such as graph neural networks or enhanced deep learning frameworks, we can leverage the underlying relationships in the data more effectively. These models, already proven successful in various scenarios, can provide more reliable predictions even in cold-start situations where traditional models often fall short due to data sparsity.

Additionally, leveraging transfer learning techniques to transfer knowledge from well-established models to new, less active models can help bootstrap the recommendation process for cold-start scenarios. Transfer learning allows the system to apply knowledge gained from one domain (e.g., existing users and items) to another domain (e.g., new users and items), thus improving the accuracy and efficiency of the recommendation process. By utilizing pre-trained models or embeddings, the system can quickly adapt to new data and provide relevant recommendations even in the absence of extensive interaction history.

Moreover, developing adaptive federated learning frameworks that can dynamically adjust the participation and contribution of clients based on their activity levels and data quality can further enhance scalability. Such frameworks can ensure that computational and communication resources are allocated efficiently, prioritizing clients that contribute more valuable information to the overall model. This dynamic adjustment can help manage the scalability challenges posed by an increasing number of clients while maintaining high-quality recommendations. Adaptive frameworks can also reduce the communication overhead by selectively updating the model parameters based on the most informative client contributions, thus optimizing the overall system performance.

Another promising direction is the incorporation of advanced trust mechanisms that can continuously evaluate and adjust the credibility of client contributions. By employing techniques such as reputation systems and anomaly detection algorithms, federated learning systems can more effectively filter out unreliable or malicious contributions, ensuring that the aggregated model remains robust and accurate. Trust mechanisms play a crucial role in maintaining the integrity of the federated learning process, especially in environments where clients may have varying levels of data quality or may be subject to adversarial attacks.

In summary, addressing cold-start issues and enhancing federated scalability are critical for the continued success and expansion of federated recommendation systems. By leveraging rich side information, incorporating credibility-based mechanisms, utilizing clustering techniques, and exploring advanced machine learning and adaptive frameworks, federated learning systems can overcome the challenges posed by large-scale participation and limited interaction data. Continued research and innovation in these areas will be essential for developing robust, scalable, and effective federated recommendation systems that can cater to a diverse and growing user base. As federated learning continues to gain traction in various industries, the ability to effectively manage cold-start issues and scalability will be key to delivering high-quality, personalized recommendations that meet the needs of users in a rapidly evolving digital landscape.

5.6 Tackling the Non-IID Issue and Personalized Federation

Non-IID (Non-Independent and Identically Distributed) data is an inevitable and challenging issue among federated models [17, 18, 45]. Before the presence of FL, RecSys had already noticed the performance degradation and divergence caused by non-IID preference data [2]. For instance, a user's behaviors may be guided by friends or groups, and adjacent items in a consecutive sequence may be related to each other. As FL confines raw data to be kept locally, extreme non-IIDness is more common than in the centralized mode, manifesting as attribute skew, label skew, and quantity skew [46]. The challenge of non-IID data in federated learning environments is exacerbated by the diverse and often conflicting data distributions across clients, making it difficult to train a globally effective model that performs well for all participants.

One effective solution to this problem is personalized federated learning, which has attracted significant attention [29]. Personalized federated learning aims to tailor the federated model to individual client needs while still benefiting from the shared global model. This approach allows clients to maintain personalized models that better align with their local data distributions while still contributing to and benefiting from the collective knowledge encoded in the global model. This dual advantage is crucial for addressing the non-IID issue, as it enables the federated system to account for the inherent variability in client data.

Personalized federated learning approaches can be categorized into data-based, model-based, and similarity-based techniques, each addressing the non-IID issue from different angles.

Data-based methods focus on strategically selecting clients that contribute the most to the model's performance. For example, client selection techniques [19] aim to prioritize clients whose data distributions are most beneficial for the global model. By selectively engaging clients with valuable data, these methods help mitigate the adverse effects of non-IID data, ensuring that the global model is trained on a representative subset of the overall data distribution. This targeted selection process reduces the likelihood of model divergence caused by conflicting data from highly heterogeneous clients, thereby enhancing the robustness and effectiveness of the federated model.

Model-based approaches involve creating models that can quickly adapt to new tasks with minimal data, thereby enhancing personalization. One such approach is meta-learning [33], where the model is trained to adapt rapidly to new tasks based on a few data points. Meta-learning frameworks, often referred to as "learning to learn," equip models with the ability to generalize from a limited amount of data, making them particularly effective in environments with highly non-IID data. By optimizing the model's adaptability, these approaches ensure that even clients with unique or highly skewed data distributions can achieve high levels of personalization without extensive retraining.

Similarity-based methods leverage the similarities among clients to improve model performance and personalization. Multi-task learning [16] is a prime example, where the model is trained to perform well on multiple related tasks by identifying and exploiting shared patterns across clients. This approach allows the federated system to maintain a balance between the global model's generalization ability and the local models' personalization needs. By clustering clients with similar data distributions and sharing knowledge among them, similarity-based methods enhance the overall effectiveness of the personalized models while reducing the impact of non-IID data.

In personalized federated learning, clients may trade off personalization and generality through a regularization term between the local model θ_{local} and the corresponding cluster model $\theta_{cluster}$ [35] (or the global model θ_{global} [49]). Some approaches combine θ_{local} with $\theta_{cluster}$ and θ_{global} via predefined weights [19–21, 32, 42], for example, $\theta_{local}^{*} = \gamma\theta_{local} + (1 - \gamma)\theta_{cluster}$. This interpolation helps in integrating the benefits of both local and global knowledge, ensuring that the personalized model remains effective across different user groups. By adjusting the balance between local and global model parameters, these methods provide a flexible approach to personalization that can be tailored to the specific needs and preferences of each client.

Beyond these vanilla model interpolation methods, FINDING [36] focuses on fine-grained personalization where the interpolation weights vary with the continuous training progress, and different neural layers have different personalized needs. This fine-tuned approach allows for more granular adjustments, catering to the specific personalization requirements of each neural layer, thus enhancing the overall recommendation quality. Fine-grained personalization is particularly important in complex models with multiple layers, where different layers may capture different aspects of user behavior and preferences. By allowing each layer to be personalized according to its specific role in the model, this approach maximizes the overall effectiveness of the personalized federated learning system.

The exploration of tackling the non-IID issue in HFedRec is still in its infancy. Incorporating more personalized techniques is a promising direction to further enhance recommendation performance. For example, personalized federated learning can incorporate advanced techniques such as transfer learning, where a pre-trained model from a source domain is adapted to a target domain, offering natural personalization properties [14]. Transfer learning enables the model to leverage existing knowledge, reducing the need for extensive retraining and thus improving efficiency and personalization. This approach is particularly valuable in federated learning environments where clients may have limited data, as it allows them to benefit from the broader knowledge captured by models trained on larger or more diverse datasets.

Furthermore, techniques like federated meta-learning can enable models to learn how to learn, providing quick adaptation capabilities to new clients or tasks. Federated meta-learning equips models with the ability to rapidly adapt to diverse and changing data distributions, ensuring sustained performance across different client environments. This adaptability is crucial for maintaining high-quality recommendations in dynamic and non-IID federated learning scenarios. By focusing on the meta-learning process itself, these methods ensure that the model remains flexible and capable of generalizing across a wide range of tasks and data distributions.

Future research should continue to explore and develop advanced methods for dealing with non-IID data and improving personalization in federated learning environments. This will be crucial for the scalability and effectiveness of federated recommendation systems. As federated learning continues to evolve, the integration of these advanced personalization techniques will play a pivotal role in addressing the challenges posed by non-IID data, ensuring that federated recommendation systems can deliver accurate and personalized recommendations at scale. The ongoing development of new algorithms and frameworks for personalized federated learning will be essential for overcoming the limitations of traditional federated learning approaches and unlocking the full potential of this powerful paradigm.

5.7 Open-Source Code

We have curated a comprehensive list of open-source projects associated with HFedRec as of February 2024. These projects represent a diverse array of implementations that leverage various programming languages, frameworks, and privacy-preserving techniques, reflecting the ongoing advancements and collaborations in the field of federated learning for recommendation systems.

In 2021, *FedRec_L* was introduced using Java and employed the factorization machine (FM) technique for privacy preservation. This project, which is focused on leveraging FM for secure federated recommendation, is available at https://csse.szu.edu.cn/staff/panwk/publications/FedRec. Another significant 2021 project, *FCMF*, developed in both Java and Python, integrates partially homomorphic encryption (PHE) and additive noise (AN) techniques to enhance security. The dual-language implementation underscores the versatility of federated learning frameworks, and its source code can be accessed at https://csse.szu.edu.cn/staff/panwk/publications/FCMF. Similarly, *FedMF*, written entirely in Python, also utilizes PHE to secure the federated learning process and is hosted at https://github.com/Di-Chai/FedMF.

Moving to 2022, the landscape saw the emergence of several key projects. *F2MF* and *FedRecAttack* are notable contributions from this year, both developed in Python using the PyTorch framework and employing AN techniques for privacy protection. These projects demonstrate the growing trend of using PyTorch for deep learning applications within federated learning, with repositories available at https://github.com/CharlieMat/FedFairRec and https://github.com/rdz98/FedRecAttack, respectively. Another 2022 project, *FMSS*, combines Java and Python, leveraging PyTorch while incorporating Secret Sharing (SS) and FM techniques. This multi-technique approach provides robust privacy protection and can be explored at https://github.com/LachlanLin/FMSS. Additionally, *ClusterAttack*, a 2023 project utilizing Python and PyTorch, focuses on federated attack scenarios, and its codebase is accessible via https://github.com/yflyl613/fedrec.

For deep learning-focused projects, the period from 2020 to 2023 witnessed several notable developments. *FedNewsRec* (2020) is one such project that utilizes the Keras framework alongside AN techniques to address the specific challenges of federated news recommendation systems. Its implementation is available at https://github.com/taoqi98/FedNewsRec. *Limited Negatives* (2021) leverages TensorFlow to enhance federated dual-encoder systems, and its code is accessible at https://git.io/federated_dual_encoder. Other significant contributions include *FedVAE* (2021), which explores variational autoencoders (VAEs) in a federated context using PyTorch, available at https://tinyurl.com/155zn06i, and *FedSeqRec* (2022) and *FedGME* (2022), which continue this trend of PyTorch-based implementations. Their respective repositories can be found at https://github.com/MuziLee-x/FedSeqRec and https://anonymous.4open.science/r/FedGME-F2F6.

Meta-learning, a critical area for personalized federated learning, is also represented in the open-source community. *MetaMF* (2020), developed in Python with PyTorch, explores the potential of meta-learning in federated matrix factorization and is accessible at https://bitbucket.org/HeavenDog/metamf/src/master. Reinforcement learning-based approaches are also gaining traction in federated learning, with projects like *Fed2-UCB* (2021) using Python and TensorFlow to implement multi-armed bandit strategies in a federated setting. Its repository is available at https://github.com/ShenGroup/FMAB. *FedIS* (2023), another TensorFlow-based project, focuses on federated item selection and can be explored at https://github.com/XuanangD/FedIS.

Further emphasizing the role of Python and PyTorch in federated learning, projects such as *PFedRec* (2023), *CPF-POI* (2023), and *HN3S* (2024) continue to build on the advancements in federated recommendation systems. These projects, which incorporate advanced techniques like collaborative filtering and personalized federated learning, are accessible at https://github.com/Zhangcx19/IJCAI-23-PFedRec, https://github.com/Leavesy/CPF-POI, and https://github.com/LukeZane118/HN3S, respectively.

Overall, Python has emerged as the dominant programming language for HFedRec projects, with frameworks like TensorFlow, PyTorch, Keras, Rectorch, and DeepCTR being widely adopted. These projects also highlight the integration of various privacy-preserving techniques, including Paillier homomorphic encryption (PHE), additive noise (AN), and Secret Sharing (SS), reflecting the diverse approaches to enhancing privacy and security in federated learning systems. The widespread use of these frameworks and techniques illustrates the evolving landscape of HFedRec implementations, where the focus is on balancing performance, scalability, and privacy in decentralized recommendation systems. As the field continues to grow, these open-source contributions will play a crucial role in driving innovation and collaboration, ensuring that federated learning technologies can be effectively deployed in real-world applications.

References

1. Bonawitz K, Ivanov V, Kreuter B, Marcedone A, McMahan HB, et al (2017) Practical secure aggregation for privacy-preserving machine learning. In Proceedings of the 2017 ACM SIGSAC conference on computer and communications security, CCS '17. ACM Press, New York, pp 1175–1191
2. Cao L (2016) Non-iid recommender systems: a review and framework of recommendation paradigm shifting. Engineering 2(2):212–224
3. Chai D, Wang L, Chen K, Yang Q (2022) Efficient federated matrix factorization against inference attacks. ACM Trans Intell Syst Technol **13**(4):1–20
4. Chen M, Zhang W, Yuan Z, Jia Y, Chen H (2022) Fede: Embedding knowledge graphs in federated setting. In Proceedings of the 10th international joint conference on knowledge graphs, IJCKG '21. Association for Computing Machinery, New York, pp 80–88
5. Cheon JH, Kim A, Kim M, Song Y (2017) Homomorphic encryption for arithmetic of approximate numbers. In Advances in Cryptology—ASIACRYPT 2017. Springer, Cham, pp. 409–437
6. Guo Q, Zhuang F, Qin C, Zhu H, Xie X, Xiong H, He Q (2022) A survey on knowledge graph-based recommender systems. IEEE Trans Knowl Data Eng 34(8):3549–3568
7. Halevi S, Polyakov Y, Shoup V (2018) An improved RNS variant of the BFV homomorphic encryption scheme. Cryptology ePrint Archive, Paper 2018/117. https://doi.org/10.1007/978-3-030-12612-4_5. https://eprint.iacr.org/2018/117
8. He X, Deng K, Wang X, Li Y, Zhang Y, Wang M (2020) Lightgcn: simplifying and powering graph convolution network for recommendation. In Proceedings of the 43rd International ACM SIGIR conference on research and development in information retrieval. ACM Press, New York, pp 639–648
9. He X, Liao L, Zhang H, Nie L, Hu X, Chua TS (2017) Neural collaborative filtering. In Proceedings of the web conference, WWW '17. International World Wide Web Conferences Steering Committee, Republic and Canton of Geneva, CHE, pp 173–182
10. Hu Y, Koren Y, Volinsky C (2008) Collaborative filtering for implicit feedback datasets. In IEEE international conference on data mining, ICDM '08. IEEE Computer Society, USA, pp 263–272
11. Huba D, Nguyen J, Malik K, Zhu R, Rabbat M, Yousefpour A, Wu CJ, Zhan H, Ustinov P, Srinivas H, et al (2022) Papaya: practical, private, and scalable federated learning. Proceedings of Machine Learning and Systems 4:814–832
12. Jiang Z, Wang W, Liu Y (2021) Flashe: additively symmetric homomorphic encryption for cross-silo federated learning. arXiv:2109.00675
13. Jordon J, Yoon J, Van Der Schaar M (2019) Pate-gan: generating synthetic data with differential privacy guarantees. In International Conference on Learning Representations
14. Kong X, Wang K, Hou M, et al (2021) A federated learning-based license plate recognition scheme for 5g-enabled internet of vehicles. IEEE Trans Industr Inform 17(12):8523–8530
15. Li D, Wang J (2019) Fedmd: heterogenous federated learning via model distillation. arXiv:1910.03581
16. Li T, Hu S, Beirami A, Smith V (2021) Ditto: fair and robust federated learning through personalization. In: Proceedings of the 38th international conference on machine learning, vol 139. PMLR, New York, pp 6357–6368
17. Li T, Sahu AK, Zaheer M, Sanjabi M, Talwalkar A, Smith V (2020) Federated optimization in heterogeneous networks. Proceedings of Machine Learning and Systems 2:429–450
18. Li X, Jiang M, Zhang X, Kamp M, Dou Q (2021) Fedbn: federated learning on non-iid features via local batch normalization. In: International Conference on Learning Representations
19. Luo S, Xiao Y, Liu Y, Li C, Song L (2022) Towards communication efficient and fair federated personalized sequential recommendation. In International conference on information communication and signal processing

20. Luo S, Xiao Y, Song L (2022) Personalized federated recommendation via joint representation learning, user clustering, and model adaptation. In Proceedings of the 31st ACM international conference on information & knowledge management, CIKM '22. Association for Computing Machinery, New York, pp 4289–4293
21. Luo S, Xiao Y, Zhang X, Liu Y, Ding W, Song L (2023) Perfedrec++: enhancing personalized federated recommendation with self-supervised pre-training. ACM Trans Intell Syst Technol
22. Paillier P (1999) Public-key cryptosystems based on composite degree residuosity classes. In: Proceedings of the 17th international conference on theory and application of cryptographic techniques, EUROCRYPT'99. Springer, Berlin, pp 223–238
23. Peng H, Li H, Song Y, Zheng V, Li J (2021) Differentially private federated knowledge graphs embedding. In Proceedings of the 30th ACM international conference on information & knowledge management. ACM Press, New York, pp 1416–1425
24. Qin Z, Deng S, Zhao M, Yan X (2023) Fedapen: personalized cross-silo federated learning with adaptability to statistical heterogeneity. In Proceedings of the 29th ACM SIGKDD conference on knowledge discovery and data mining, KDD '23. Association for Computing Machinery, New York, pp 1954–1964
25. Rendle S, Freudenthaler C, Gantner Z, Schmidt-Thieme L (2009) BPR: Bayesian personalized ranking from implicit feedback. In: Proceedings of the twenty-fifth conference on uncertainty in artificial intelligence, UAI '09. AUAI Press, Arlington, pp 452–461
26. Rong D, Ye S, Zhao R, Yuen HN, Chen J, He Q (2022) Fedrecattack: model poisoning attack to federated recommendation. In IEEE 38th international conference on data engineering. IEEE Computer Society, Los Alamitos, pp 2643–2655
27. Salakhutdinov R, Mnih A (2007) Probabilistic matrix factorization. In Proceedings of the 20th international conference on neural information processing systems, NIPS'07. Curran Associates Inc., Red Hook, pp 1257–1264
28. Shen T, Zhang J, Jia X, Zhang F, Huang G, Zhou P, Kuang K, Wu F, Wu C (2020) Federated mutual learning. arXiv:2006.16765
29. Tan AZ, Yu H, Cui L, Yang Q (2023) Towards personalized federated learning. IEEE Trans Neural Networks Learn Syst 34(12):9587–9603
30. Tan Y, Long G, Liu L, Zhou T, Lu Q, Jiang J, Zhang C (2022) Fedproto: federated prototype learning across heterogeneous clients. In Proceedings of the AAAI conference on artificial intelligence, pp 8432–8440
31. Wahab OA, Rjoub G, Bentahar J, Cohen R: Federated against the cold: a trust-based federated learning approach to counter the cold start problem in recommendation systems. Inf Sci 601, 189–206 (2022)
32. Wang S, Tao H, Li J, Ji X, Gao Y, Gong M (2024) Towards fair and personalized federated recommendation. Pattern Recogn 149:110234
33. Yang L, Huang J, Lin W, Cao J (2023) Personalized federated learning on non-iid data via group-based meta-learning. ACM Trans Knowl Discov Data 17(4):1–20
34. Ye M, Fang X, Du B, Yuen PC, Tao D (2023) Heterogeneous federated learning: state-of-the-art and research challenges. ACM Comput Surv 56(3):1–44
35. Ye Z, Zhang X, Chen X, Xiong H, Yu D (2023) Adaptive clustering based personalized federated learning framework for next poi recommendation with location noise. IEEE Trans Knowl Data Eng 36(5):1–1
36. Yu SL, Liu Q, Wang F, Yu Y, Chen E (2023) Federated news recommendation with fine-grained interpolation and dynamic clustering. In Proceedings of the 32nd ACM international conference on information and knowledge management, CIKM '23. Association for Computing Machinery, New York, pp 3073–3082
37. Yuan W, Qu L, Cui L, Tong Y, Zhou X, Yin H (2024) Hetefedrec: federated recommender systems with model heterogeneity. In: IEEE 40th International Conference on Data Engineering
38. Yuan W, Yang C, Nguyen QVH, Cui L, He T, Yin H (2023) Interaction-level membership inference attack against federated recommender systems. In Proceedings of the Web Conference, WWW '23. Association for Computing Machinery, New York, pp 1053–1062

39. Zhang C, Li S, Xia J, Wang W, Yan F, Liu Y (2020) Batchcrypt: efficient homomorphic encryption for cross-silo federated learning. In Proceedings of the 2020 USENIX Conference on Usenix Annual Technical Conference. USENIX Association, USA
40. Zhang C, Long G, Zhou T, Zhao X, Zhang Z, Yang B (2023) When Federated Recommendation Meets Cold-Start Problem: Separating Item Attributes and User Interactions. arXiv:2305.12650
41. Zhang M, Wei E, Berry R (2021) Faithful edge federated learning: scalability and privacy. IEEE J Sel Areas Commun 39(12):3790–3804
42. Zhang X, Ye Z, Lu J, Zhuang F, Zheng Y, Yu D (2023) Fine-grained preference-aware personalized federated poi recommendation with data sparsity. In Proceedings of the 46th International ACM SIGIR conference on research and development in information retrieval, SIGIR '23. Association for Computing Machinery, New York, pp 413–422
43. Zhao WX, Hou Y, Pan X, Yang C, Zhang Z, Lin Z, Zhang J, Bian S., Tang J, Sun W, Chen Y, Xu L, Zhang G, Tian Z, Tian C, Mu S, Fan X, Chen X, Wen JR (2022) Recbole 2.0: Towards a more up-to-date recommendation library. In Proceedings of the 31st ACM international conference on information & knowledge management, CIKM '22. Association for Computing Machinery, New York, pp 4722–4726
44. Zhao WX, Mu S, Hou Y, Lin Z, Chen Y, Pan X, Li K, Lu Y, Wang H, Tian C, Min Y, Feng Z, Fan X, Chen X, Wang P, Ji W, Li Y, Wang X, Wen JR (2021) Recbole: towards a unified, comprehensive and efficient framework for recommendation algorithms. In Proceedings of the 30th ACM international conference on information & knowledge management, CIKM '21. Association for Computing Machinery, New York, pp 4653–4664
45. Zhao Y, Li M, Lai L, Suda N, Civin D, Chandra V (2018) Federated learning with non-iid data. arXiv:1806.00582
46. Zhu H, Xu J, Liu S, Jin Y (2021) Federated learning on non-iid data: a survey. Neurocomputing 465:371–390
47. Zhu J, Dai Q, Su L, Ma R, Liu J, Cai G, Xiao X, Zhang R (2022) Bars: towards open benchmarking for recommender systems. In Proceedings of the 45th International ACM SIGIR conference on research and development in information retrieval, SIGIR '22. Association for Computing Machinery, New York, pp 2912–2923
48. Zhu J, Liu J, Yang S, Zhang Q, He X (2021) Open benchmarking for click-through rate prediction. In Proceedings of the 30th ACM international conference on information & knowledge management, CIKM '21. Association for Computing Machinery, New York, pp 2759–2769
49. Zhu Z, Si S, Wang J, Xiao J (2022) Cali3f: calibrated fast fair federated recommendation system. In International Joint Conference on Neural Networks
50. Zoph B (2016) Neural architecture search with reinforcement learning. arXiv preprint arXiv:1611.01578

The manufacturer's authorised representative in the EU is Springer Nature Customer Service Centre GmbH, Europaplatz 3, 69115 Heidelberg, Germany. If you have any concerns regarding our products, please contact ProductSafety@springernature.com

Printed and bound by CPI Group (UK) Ltd, Croydon, CR0 4YY
01/10/2026
02225944-0014